77th Annual Meeting of Sugar Industry Technologists 2018

AA005642

Sugar Industry Technologists, Inc. Volume LXXVII

Bonita Springs, Florida, USA
25 - 28 March 2018

ISBN: 978-1-5108-7017-8

Printed from e-media with permission by:

Curran Associates, Inc.
57 Morehouse Lane
Red Hook, NY 12571

Some format issues inherent in the e-media version may also appear in this print version.

Printed by Curran Associates, Inc. (2018)

For permission requests, please contact Sugar Industry Technologists Inc. (SIT)
at the address below.

Sugar Industry Technologists Inc. (SIT)
4032 Captiva Row
Myrtle Beach, South Carolina 29579
USA

Phone: +1 863 983 3637
Fax: +1 843 796 4743

SITcontact@sucrose.com

Additional copies of this publication are available from:

Curran Associates, Inc.
57 Morehouse Lane
Red Hook, NY 12571 USA
Phone: 845-758-0400
Fax: 845-758-2633
Email: curran@proceedings.com
Web: www.proceedings.com

TABLE OF CONTENTS

TECHNICAL PRESENTATIONS

POSTERS:

Roster of Delegates

Roster of Guests

SUGAR INDUSTRY TECHNOLOGISTS, INC.

VOLUME LXXVII

Publication of Technical Papers and Proceedings

of the

SEVENTY SEVENTH ANNUAL MEETING

of

SUGAR INDUSTRY TECHNOLOGISTS, INC.

BONITA SPRINGS, FLORIDA, U.S.A.
MARCH 25 – 28, 2018

Published by

SUGAR INDUSTRY TECHNOLOGISTS, INC.

New York

Mailing Address
4032 Captiva Row
Myrtle Beach, South Carolina 29579, U.S.A.
WEB SITE: https://sugarindustrytechnologists.com

PURPOSE OF SUGAR INDUSTRY TECHNOLOGISTS, INC.

Sugar Industry Technologists, Inc. was established in 1941 to serve the professional interests of its members by providing a forum for the exchange of scientific information and other technical aspects of sugar refining, by encouraging original research in sugar technology, by promoting a broader acquaintance among its members and by cooperating with other engineering, technical and scientific societies.

Past Meeting Sites

1941 - 1967, New York
1968 - Montreal, Canada
1969 - New York
1970 - New Orleans
1971 - New York
1972 – Houston
1973 - Miami
1974 - Vancouver, Canada
1975 - Savannah
1976 - Toronto, Canada
1977 - San Francisco
1978 - London, England
1979 - Boston
1980 - Chicago
1981 - Fort Lauderdale
1982 - Atlanta
1983 - New York
1984 - Houston
1985 - Saint John, Canada
1986 - Baltimore
1987 – Sydney, Australia
1988 – Savannah, GA, USA
1989 - New Orleans, LA, USA
1990 - Vancouver, Canada
1991 - New York
1992 – Brussels, Belgium
1993 – Toronto, Canada
1994 - Honolulu, Hawaii
1995 - La Romana, Dominican Republic
1996 - Durban, South Africa
1997 - Montreal, Canada
1998 - Marseille, France
1999 - Estoril, Portugal
2000 - New Orleans
2001 –Taipei, Taiwan, ROC
2002 - Delray Beach, Florida
2003 – Hamilton Island, Australia
2004 – Vancouver, Canada
2005 – Dubai, United Arab Emirates
2006 – La Baule, France
2007 – Baltimore, Maryland, USA
2008 – Leipzig, Germany
2009 – New Orleans, LA, USA
2010 – Savannah, GA, USA
2011 – Montreal, Canada
2012 – Auckland, New Zealand
2013 - Guangzhou, China
2014 - Toronto, Canada
2015 – Osaka, Japan
2016 - New York, NY, USA
2017 – Tainan, Taiwan, R.O.C.
2018 – Bonita Springs, LA, USA

APPROVED FUTURE MEETING SITES

2019 – May 5-8, Durban, South Africa
2020 – May 3-6, New Orleans, LA, USA

FOREWORD

The 77th Annual Technical Meeting of the Sugar Industry Technologists, Inc. was held in Bonita Springs, Florida, U.S.A. from March 25th to March 28th. Eighteen countries were represented and included 158 delegates and 39 guests for a total of 197 attendees.

The guest speaker was Mr. Robert H Buker, the President and CEO of the United States Sugar Corporation. The theme of his presentation was Operations Management.

Four technical sessions were presided over by Michael Burchell (ASR Group – C&H Sugar). David Demone (ASR Group – C&H Sugar), Ahmed Vawda (United Sugar Company) and Jack Thompson (Louisiana Sugar Refining LLC). Thirteen technical papers were presented and all were well received with numerous questions from the audience. Two high quality posters were on display for viewing between sessions. Many thanks to Michael Burchell for organising the very successful technical program.

Two symposiums were held, the topics "Sustainability" and "Improving Yield" providing much interest as evidenced by the amount of discussion and questions addressed to the panelists. The panels had a good mix of factory operations staff, equipment suppliers and research organisations allowing for a balanced perspective on the issues discussed. Thanks to Jack Thompson for organizing and moderating the symposiums.

On March 28th the delegates and guests visited the United States Sugar Corporation in Clewiston, and the trip included visits to cane fields, the raw sugar factory as well as the Refinery! Thanks to United States Sugar Corporation for arranging the very extensive visits, their hospitality during these tours and the luncheon.

Special thanks to Michael Burchell and Jack Thompson for standing in for me for the Opening, presiding over the Banquet Ceremony and the closing speech.

As always, a further special thank you goes to Edgar Aguirre, Executive Director, for all his efforts to make this another smooth, successful and professional technical conference.

As the incoming S.I.T. President I would like to personally invite you to the 78th S.I.T. Technical Meeting on the 5th to 8th of May in Durban, South Africa.

Paul Schorn
2018-2019 S.I.T. President

SUGAR INDUSTRY TECHNOLOGISTS, INC.

2017-2018

OFFICERS

President …………………………..	Trevor Ohlson …………………….	Lantic Inc. - Vancouver
1st Vice President ……………….	Paul Schorn …………………….…	Tongaat Hulett Sugar South Africa Ltd.
2nd Vice President ……………...	Jack Thompson……………….….	Louisiana Sugar Refining, LLC
Vice President – Finance.……...	David Demone …………………..	ASR Group - American Sugar Refining, Inc.
Immediate Past President ……..	Michael Burchell ……………..….	ASR Group – C&H Sugar Company, Inc.
Executive Director.....................	Edgar Aguirre	Sugar Industry Technologists, Inc.

EXECUTIVE COMMITTEE

Sune Berghall ………………………………………………………	Finnsugar Ltd.
David Meadows	Tongaat Hulett Ltd.
David Pelham	United States Sugar Corporation
Thomas Rathke ………………………………………………....	Imperial Sugar Company
Frank van Noord ………………………………………………....	Suiker Unie
Katarina Silfversparre……………………………………………….	Nordic Sugar AB
Peter Simpson……………………………………………………….	New Zealand Sugar Co. Ltd.
Ahmed Vawda……………………………………………………….	United Sugar Company

COMMITTEE CHAIRMAN

By-Laws....................................	Thomas Rathke	Imperial Sugar Company
Finance ………………………..	David Demone ……………….	ASR Group - American Sugar Refining, Inc.
Crystal Award ………………..	Peter Maraia…..………………...	ASR Group - American Sugar Refining, Inc.
Meade Award ………………...	Paul Schorn ………………..……...	Tongaat Hulett Sugar South Africa Ltd.
Chapman Memorial Award ….	Paul Schorn ………………..……...	Tongaat Hulett Sugar South Africa Ltd.
Nominating ………………….	Ahmed Vawda ………………….	United Sugar Company
Membership…………………….	Edgar Aguirre …………………..	Sugar Industry Technologists, Inc.
Symposium ………………….	Jack Thompson ……………..……..	Louisiana Sugar Refining, LLC
Technical Meeting Program…..	Michael Burchell……………….	ASR Group – C&H Sugar Company, Inc.
Publications..............................	Edgar Aguirre..............................	Sugar Industry Technologists, Inc.

LOCAL ARRANGEMENTS

2018 – Bonita Springs, FL, U.S.A.	David Pelham ………………….	United States Sugar Corporation
2019 – Durban, South Africa	Paul Schorn ………………..……...	Tongaat Hulett Sugar South Africa Ltd.

S.I.T. OFFICERS

2017 - 2018

Edgar Aguirre, Jack Thompson, David Demone and Michael Burchell
Not Shown: Trevor Ohlson and Paul Schorn

PAUL SCHORN
President-Elect

Jack Thompson
1st VP Elect

Jamal Abdel Kaddoury
2nd VP Elect

SUGAR INDUSTRY TECHNOLOGISTS, INC.

2018 – 2019

OFFICERS

President	Paul Schorn	Tongaat Hulett Sugar South Africa Ltd.
1st Vice President	Jack Thompson....................	Louisiana Sugar Refining, LLC
2nd Vice President	Abdel Jalil Kaddoury..............	COSUMAR S.A.
Vice President – Finance	David Demone	ASR Group - American Sugar Refining, Inc.
Immediate Past President	Michael Burchell..................	ASR Group – C&H Sugar Company, Inc.
Executive Director..................	Edgar Aguirre	Sugar Industry Technologists, Inc.

EXECUTIVE COMMITTEE

Sune Berghall..	Finnsugar Ltd.
David Pelham	United States Sugar Corporation.
Dave Meadows..	Tongaat Hulett Ltd.
Frank van Noord	Suiker Unie
Thomas Rathke	Imperial Sugar Company
Katarina Silfversparre..	Nordic Sugar AB
Peter Simpson..	New Zealand Sugar Co., Ltd.
Ahmed Vawda..	United Sugar Company

COMMITTEE CHAIRMAN

By-Laws..................................	Thomas Rathke..........................	Imperial Sugar Company
Finance	David Demone	ASR Group - American Sugar Refining, Inc.
Crystal Award	David Demone	ASR Group - American Sugar Refining, Inc.
Meade Award	Jack Thompson	Louisiana Sugar Refining, LLC
Chapman Memorial Award	Jack Thompson	Louisiana Sugar Refining, LLC
Nominating	Ahmed Vawda	United Sugar Company
Membership......................	Edgar Aguirre......................	Sugar Industry Technologists, Inc.
Symposium	Abdel Jalil Kaddoury..............	COSUMAR S.A.
Technical Meeting Program.....	Michael Burchell..................	ASR Group – C&H Sugar Company, Inc.
Publications	Edgar Aguirre..........................	Sugar Industry Technologists, Inc.

LOCAL ARRANGEMENTS

2019 – Durban, South Africa	Paul Schorn	Tongaat Hulett Sugar South Africa Ltd.
2020 – New Orleans, LA, USA	Jack Thompson	Louisiana Sugar Refining, LLC

SUGAR INDUSTRY TECHNOLOGISTS, INC.

DIRECTORS

2018-2019

PAST PRESIDENTS

of

SUGAR INDUSTRY TECHNOLOGISTS, INC.

1941 **ROBERT S. KENT***
Robert Sayre Kent, Inc.

1942 **ROBERT S. KENT***
Robert Sayre Kent, Inc.

1943 **R. A. RING**
Atlantic Sugar Refineries Co. Ltd.

1944 **ALEX M. ORMOND**
Savannah Sugar Refining Corporation

1945 **FRED M. GIBSON**
American Sugar Refining Company

1946 **FRANK C. STAPLES**
American Molasses Company

1947 **FRANK C. STAPLES**
American Molasses Company

1948 **FRANK C. STAPLES**
American Molasses Company

1949 **FRANK M. EXLEY**
Savannah Sugar Refining Company

1950 **CURTIS L. TAGGART**
Revere Sugar Refinery

1951 **ANSON J. FOWLER**
National Sugar Refining Company

1952 **JAMES M. BROWN**
Revere Sugar Refinery

1953 **JAMES M. BROWN**
Revere Sugar Refinery

1954 **FRANK M. CHAPMAN**
Tate & Lyle Refineries, Ltd.

1955 **PETER H. PETRI**
American Molasses Company

1956 **GEOFFREY CONSTABLE**
St. Lawrence Sugar Refineries, Ltd.

1957 **FRANK A. MONROE, JR.**
Revere Sugar Refinery

1958 **W. AUSTIN CLARNO**
Savannah Sugar Refining Company

1959 **ELMER J. CULP**
American Sugar Refining Company

1960 **G. ERNEST WARING**
Atlantic Sugar Refining Co., Ltd.

1961 **A. D. HINRICHS**
Refined Syrups & Sugar Inc.

1962 **H. J. JACOBS**
South Coast Corporation

1963 **FREDERICK BRUDER**
SuCrest Corporation

1964 **WATSON A. BEMIS**
Revere Sugar Refinery

1965 **FREDERICK W. SCHWER**
National Sugar Refining Company

1966 **E. D. GILLETTE**
Refined Syrups & Sugars, Inc

1967 **RUFUS J. HERRING**
Savannah Sugar Refining Corporation

1968 **H. JAMES REICHE**
Refined Syrups & Sugars Inc.

1969 **HENRY G. GERSTNER**
Colonial Sugars Company

1970 **DORR E. TIPPENS**
Amstar Corporation

1971 **PHILIP F. MEADS**
California & Hawaiian Sugar Co.

1972 **JOSEPH A. METZLER**
Goudchaux-Henderson Sugar Co.,

**Chairman of Executive Committee*

PAST PRESIDENTS

of

SUGAR INDUSTRY TECHNLOLGISTS, INC.

1973 **WARREN L. REED**
Revere Sugar Refinery

1974 **STEPHEN STACHENKO**
Redpath Sugars Limited

1975 **GEORGE FAWCETT**
Savannah Sugar Refinery

1976 **A. M. BARTOLO**
Imperial Sugar Company

1977 **K. R. HANSON**
Amstar Corp. American Sugar Div.

1978 **STANLEY E. GEORGE**
B. C. Sugar

1979 **R. STUART PATTERSON**
California & Hawaiian Sugar Co.

1980 **JOHN V. LOPEZ-ONA**
National Sugar Refining Company

1981 **JOSEPH A. HARRISON, SR.**
Supreme Sugar Company

1982 **MICHAEL C. BENNETT**
Tate & Lyle, Ltd.

1983 **BRUCE A. FOSTER**
St. Lawrence Sugar, Ltd.

1984 **LOUIS E. MAHONEY III**
Revere Sugar Corporation

1985 **EDWARD D. STEPHENSON**
Lantic Sugar Limited

1986 **DAVID M. HUMM**
California & Hawaiian Sugar Co.

1987 **LEON A. ANHAISER**
Imperial Holly Sugar Company

1988 **C. FRANK STOWE**
Amstar Sugar Corporation

1989 **LEON SUE**
Raffinerie Tirlemontoise S. A...

1990 **CHRISTIAN LAUR**
Tate & Lyle France

1991 **JOSEPH W. ALBERINO**
Savannah Sugar Refinery

1992 **THOMAS F. CHANDLER**
Redpath Sugars, Limited

1993 **DONALD E. WEBSTER**
B. C. Sugar

1994 **BRIAN T. HARRISON**
Imperial Holly Corporation

1995 **C. RICHARD DONNELLY**
Everglades Sugar Refinery, Inc.

1996 **KEITH SINCLAIR**
Tate & Lyle Sugars

1997 **MARCEL BRAECKMAN**
Raffinerie Tirlemontoise, S. A.

1998 **LUIS SAN MIGUEL BENTO**
Refinarias de Acucar Reunidas

1999 **RAY BURKE**
Savannah Sugar Refinery

1999 **LUIS SAN MIGUEL BENTO**
Refinarias de Acucar Reunidas

2000 **CHUNG CHI CHOU**
Dr. Chou Technologies, Inc.

2001 **STEPHEN CLARKE**
Florida Crystals Corporation.

2002 **DENIS BOURÉE**
Saint Louis Sucre

2003 **JEAN-PAUL MERLE**
C & H Sugar Company

2004 **EDGAR AGUIRRE**
2005 *United States Sugar Corp*

2006 **JEFFREY ROBINSON**
American Sugar Refining Inc.

2007 **PHILIP ROWLAND**
Imperial Sugar Company

2008 **DONALD RYAN**
Lantic Inc. – Vancouver

PAST PRESIDENTS

of

SUGAR INDUSTRY TECHNOLOGISTS, INC.

2009 **AMELIA FERRAZ**
Rar-Refinarias de Azucar Reunides

2010 **AHMED VAWDA**
United Sugar Company

2011 **DAVID DEMONE**
ASR Group - American Sugar Refining Inc.

2012 **THOMAS RATHKE**
Imperial Sugar Company

2013 **MICHAEL BURCHELL**
ASR Group – C&H Sugar Company

2014 **DAVE MEADOWS**
Tongaat Hulett Limited

2015 **DAVID PELHAM**
United States Sugar Corporation

2016 **MICHAEL BURCHELL**
ASR Group – C&H Sugar Company

2017 **TREVOR OHLSON**
Lantic Inc. - Vancouver

THE JOE HARRISON MEMORIAL SCHOLARSHIP AWARD - 2017

Awarded to Mr. Allan Maravilla, Victorias Milling Company, Philippines, for achieving the highest score at the Cane Sugar Refiner's Institute at Nicholls State University in 2017

Leon A. Anhaiser, Sugar Industry Technologists, Inc., Allan Maravilla, Victorias Milling Company, Philippines, and Jason Eiermann, Nicholls State University

FRANK CHAPMAN MEMORIAL AWARD

RECIPIENTS

2009 **GILLIAN EGGLESTON**
USDA-ARS-SRRC
BELISARIO MONTES
Alma Plantation LLC

2010 **LUIS SAN MIGUEL BENTO**
Porto

2011 **NIKLAS BRAUN**
Nordic Sugar

2012 **I. KADDOUR, A. FERREIRA, L. BENTO and F. ROCHA**
University of Porto

2013 **G. EGGLESTON, et. al.**
USDA-ARS-SRRC
GILBERT, I. ROSE
AgroParisTech

2014 **JUAN J MENDOZA, REGINA VISO,**
Prodek Inc. .
HERNANDO AFANADOR, *Ingenio Riopaila, Colombia*, **and DOMINGO SERPA,** *LIPESA*

2015 **ITSUKI SHIMASATO**
Toyo Jidoki Co., Ltd.

2016 **JOSE M. CHORAO**
RAR-Rafinarias de A úcares Reunidas
CARLOS A. ALVES and ARNALDO O. ARAÚJO
Air Liquide Portugal

2017 **MARSHA COLE, et. al.**
USDA-ARS-SRRC

JOE HARRISON MEMORIAL SCHOLARSHIP

RECIPIENTS

2004 **TERRY TAHAL**
Guyana Sugar Corporation, Guyana

2005 **RAFAEL ZAUN**
Compañía Azucarera Salvadoreña

2006 **ALBERT KATRYAN**
Guyana Sugar Corporation, Guyana

2007 **NORMA SANCHEZ FUENTES**
Sugar Saenz Group

2008 **JUAN RAMIRO SAYAGO**
Ledesma S.A.A.I., Argentina

2009 **BENJAMIN LARA**
Central Izalco, El Salvador

2010 **ADAN ERNESTO ANAYA**
Compañía Azucarera Salvadoreña

2011 **ALVARO MOISES**
Compañía Azucarera Salvadoreña

2012 **GIOVANNI MOLINA**
Compañía Azucarera Salvadoreña

2013 **DAVID ARMANDO FLAMENCO**
C.A.S.S.A. Group, El Salvador

2014 **BYRON LOPEZ**
CENGICAÑA, Guatemala

2015 **AMILCAR EDUARDO AYALA**
Ingenio La Union, Guatemala

2016 **RAMIRO R. CIFUENTES DE LEON**
Ingenio La Union, Guatemala

2017 **ALLAN MARAVILLA**
Victorias Milling Co., Philippines

THE FRANK CHAPMAN MEMORIAL AWARD - 2017

Poster No. 1174 - Development of a Factory/Refinery Method to Measure Total, Soluble, and Insoluble Starch in Sugar Products.

By

Dr. Marsha R. Cole, Dr. Gillian Eggleston, and Deriesha Gaines, USDA-ARS-SRRC
New Orleans, Louisiana, U.S.A.

Jack Thompson presenting the plaque to
Dr. Marsha R. Cole also accepting for Dr. Gillian Eggleston and Deriesha Gaines

THE GEORGE AND ELEANORE MEADE AWARD - 2017

Paper No. 1153 – Development of a Linear Pan Growth Approach

By

Chris Mayhew, Carlton Haynes, Robert Howe, Craig Parker and Chris Rhoten, British Sugar and ESCON America Inc.

Jack Thompson presenting the George and Eleanore Meade Award

To

Craig Parker, also accepting for Chris Mayhew, Carlton Haynes, Robert Howe, and Chris Rhoten

THE GEORGE AND ELEANORE MEADE AWARD

RECIPIENTS

1965 **WARREN L. REED & JAMES VLADYKA**
Revere Sugar Refinery

1966 **W. A. BENNETT; E.W.P. CUNNEEN & DR. A. M. HERTZBERG**
Colonial Sugar Refining Co. Ltd.

1967 **DR. JAMES C. P. CHEN**
Empresa Agricola Chicana, Ltd.

1968 **JOSEPH F. DOWLING**
Refined Syrups & Sugars, Inc.

1969 **A. M. MOULT**
Tate & Lyle Refineries, Ltd.

1970 **M. A. COHEN; O. G. DIONISIO & S. J. DRESHER**
American Sugar Company
DR. C. M. BENNETT & J. C. ABRAM
Tate & Lyle Technical Services

1971 **DR. CHUNG CHI CHOU & K.R. HANSON**
Amstar Corporation

1972 **P. J. LANGLEY & C. J. BOHLIG**
California & Hawaiian Sugar Co.

1973 **A. E. NORCOTT; C. V. RICH & J. T. RUNDELL**
Tate & Lyle Enterprises, Ltd.

1974 **R. S. PATTERSON**
California & Hawaiian Sugar Co.

1975 **PETER H. PETRI**
Goudchaux-Henderson Sugar Co. Ltd.

1976 **JOHN V. LOPEZ-ONA & EVERETT G. JONES**
The National Sugar Refining Co.

1977 **WILLIAM F. BARTON**
Atlantic Sugar Limited

1978 **DORR E. TIPPENS; LEIGH C. HAYES & DR. CHUNG CHI CHOU**
Amstar Sugar Corporation

1979 **WARREN L. REED**
Revere Sugar Corporation

1980 **JOHN V. LOPEZ-ONA**
The National Sugar Refining Co.

1981 **D. J. BARDWELL; H. R. DELANEY & D. T. HAWKINS**
CSR, Ltd. Sugar Division
H. C. WILSON & JOHN A. RICHMOND
Holly Sugar Corporation

1982 **W. J. BARTON**
Atlantic Sugar Limited, N.B.
W. J. KNEBEL
Calgon Corporation, Pittsburgh, PA

1983 **J. CUEL & C. LONGUE EPEE**
Begin-Say

1984 **LIEF RAMM-SCHMIDT & GORAN HYOKY**
Finnish Sugar Company, Ltd.

1985 **DIETRICH BOSSE**
BMA/Silver Engineering

1986 **DIETER FRANK; JOHN PARK & LINCOLN D. METCALFE**
Kazoo Chemise America

1987 **D. J. RADFORD; D. J. TAYFIELD & M. G. S. COX**
Hulett Refineries Limited

1988 **M. KAWAKAMI; H. IWABE & T. NAKAMURA**
Mitsui Sugar Company, Ltd.

1989 **LUIS ROCHA SAN MIGUEL BENTO**
RAR - Refinarias de Acucar Reunidas, S.A.R.L.

1990 **SIMON J. GIBBONS & RICHARD B. RATCLIFFE**
Tate & Lyle Sugars

1991 **PETER J. BILLING & MICHAEL LOW**
Tate & Lyle Sugars
GEORGE RAPOS
Redpath Sugars Limited

1992 **MARGARET A. CLARKE; LESLIE A. EDYE & NGHIA TO**
Sugar Processing Research Inst.
CYNTHIA McDONALD-LEWIS
N.I.R. Systems

THE GEORGE AND ELEANORE MEADE AWARD

RECIPIENTS

1993 **SEBASTIAO FEYO de AZEVEDO**
JOSE CHORAO, MARIA JOAO GONCALVES
University of Porto, Portugal &
LUIS SAN MIGUEL BENTO
RAR Refinarias de Acucar Reunidas S.A.

1994 **DAVE M. MEADOWS**
Tongaat-Hulett Sugar Limited

1995 **BRYAN C. TUNGLAND, GENE SNOOK, ANNETTE CAMPBELL & JANIS KARDATZKE**
Imperial Holly Corporation

1996 **GAVIN B. DALGLEISH & GRANT M. HUBBARD**
Tongaat-Hulett Sugar Limited

1997 **DERRALD H. HOUSTON & TED D. MILNER**
Silver Weibull

1998 **PETER BURNS & PETER FIELDS**
CSR Sugar

1999 **D. E. REARICK, M. KEARNEY & D. COSTESSO**
Amalgamated Research Inc.

2000 **PETER J. DIBELLA**
Manildra Harwood Sugars

2001 **MARY AN GODSHALL & KHALID IQBAL**
Sugar Processing Research Institute, Inc.

2002 **JAMAL AL GHURAIR, G. C. SINGH, & B. K. PATIL**
Al Khaleej Sugar Company

2003 **P. SIMPSON, A. CARREON & S. SPILIOPOULOS**
New Zealand Sugar Company & Sugar Australia

2004 **PHILIP ROWLAND**
Imperial Sugar Company – Savannah

2005 **REINHOLD HEMPELMANN**
BMA

2006 **LES EDYE**
Queensland University
GAVIN KHAN
Carbochem Inc.

2007 **ARNO JANSEN**
Schaffer and Associates
MALCOLM WADE
United States Sugar Corporation

2008 **GEORGE CARTER**
Redpath Sugar

2009 **BARBARA MUIR, SMRI, and**
GILLIAN EGGLESTON, USDA-ARS-SRRC

2010 **AHMED VAWDA AND G.N. KASSIM**
Savola Sugar Middle East, Cairo, Egypt

2011 **GEORGE CARTER**
Redpath Sugar

2012 **JEFFREY C. ROBINSON**
American Sugar Refining, Inc.

2013 **PETER SIMPSON**
New Zealand Sugar Co. Ltd.

2014 **MARSHA R. COLE, GILLIAN EGGLESTON**
YOO JIN CHUNG, ***USDA-ARS-SRRC and***
AUDREY GILBERT, ***AgroParisTech***

2015 **CRAIG JENSEN & DAVID LOVE**
Tongaat Hulett Limited

2016 **JESSE WILLIAMS & ELIZABETH MENDONCA**
ASR Group, American Sugar Refining, Inc.

2017 **CHRIS MAYHEW, CARLTON HAYNES, ROBERT HOWE, CRAIG PARKER, AND CHRIS RHOTEN,** ***British Sugar and Escon America Inc.***

S.I.T. CRYSTAL AWARD 2018

David P. Demone

Dave was born in Massachusetts in 1949 and attended Northeastern University in Boston. His work-study position was as a laboratory analyst for the American Sugar Refining Company's Bunker Hill Refinery, producing the well-known Domino Sugar brand, beginning in June of 1968. This small "experimental" refinery had a 2 million pound per day capacity and a culture that encouraged direct contact with all levels of operations and management. Learning the cane sugar refining process from the perspective of analysis of process streams and products was immediately fascinating and ultimately proved to be an invaluable experience.

After graduating with a Chemical Engineering degree in 1972, Dave accepted a position as Plant Chemist. Within a year Dave was transferred to a position as Utilities / Environmental Engineer and the following year moved to the Process Department as an Area Engineer. In 1979, Dave relocated to the Baltimore Refinery as a Process Engineer to expand his process knowledge. The Boston plant had no traditional char house, soft sugar system or Sweetland presses. In 1984, there was an opportunity to work in the Manhattan corporate office so the family moved to New Jersey. This assignment was short-lived and after five months Dave became Process Manager at the Brooklyn Refinery and remained in Brooklyn until the refinery ceased operation in 2004. From 1990 until the closure of the refinery Dave was the Environmental & Quality Control Manager. From 2004 until present Dave has been a member of the corporate quality team working at the Yonkers Refinery and for the last 10 years at the corporate sales office in Iselin, NJ.

In 1992, the corporation took a bold step forward and embarked on an aggressive program to be the first in the industry to achieve ISO 9002 certification. As the Environmental & Quality Manager Dave was the Brooklyn representative and soon became the corporate representative traveling to all sites for audits and interfacing with the certifying body. The significant involvement in quality systems and certifications has remained constant over the past 25 years. Additional current responsibilities include assisting in processing and technical issues at all location, providing support to customers and acting as a liaison between sales, financial and operations. His current position is Quality Assurance Programs Manager for ASR Group, American Sugar Refining, Inc.

The nearly 50-year journey as a cane sugar refiner, beginning with a co-operative work-study position in his field and culminating in a lifelong career, is amazing in retrospect.

Dave's involvement with SIT began in 2004. He has been a member of the Board of Directors and Executive Committee. Dave served as President in 2012 and is currently the Vice President - Finance. The friendships gained and amazing experiences resulting directly from SIT participation will last a lifetime.

Dave and Kathy have been married for 47 years. They have a daughter, Laura, a son, Mark, a daughter-in-law, Cristina and six wonderful grandchildren, Erik, Matthew, Carina, Luke, Jack and Nicholas. He is grateful for the support and encouragement provided by his family throughout his career.

S.I.T. CRYSTAL AWARD - 2018

Michael Burchell presents the Crystal Award

To

David P. Demone

RECIPIENTS OF THE S.I.T. CRYSTAL AWARD**

(S.I.T. Achievement Award in Sugar Technology)

1961 **DR. VICTOR DIETZ**
National Bureau of Standards

1961 **LOHN LOWE**
Revere Sugar Refinery

1962 **HAROLD E. C. POWERS**
Tate & Lyle Refineries, Ltd.

1962 **EUGENE C. GILLETT**
C & H Sugar Refining Corporation

1963 **DR. GEORGE P. MEADE**
Colonial Sugars Company

1963 **ALFRED L. WEBRE**
Jackson Industries, Inc.

1964 **FRANK M. EXLEY**
Savannah Sugar Refining Corp.

1964 **HAROLD J. JACOBS. SR.**
The South Coast Corp.

1965 **FRANK M. CHAPMAN**
Tate & Lyle Refineries, Ltd.

1965 **DR. PIETER HONIG**
Consultant

1966 **WILLIAM H LOUVIERE**
Imperial Sugar Company

1966 **PAUL P. WOEHRLE**
National Sugar Refining Co.

1967 **DR. ELLIOT P. BARRETT**
Kerr-McGee Chemical Corp.

1967 **WATSON A. BEMIS**
Revere Sugar Refinery

1968 **E. DWIGHT GILLETTE**
Refined Syrups & Sugars, Inc.

1969 **DR. PHILIP F. MEADS**
California & Hawaiian Sugar Co.

1969 **ELMER J. CULP**
American Sugar Company

1970 **HENRY G. GERSTNER**
Colonial Sugars Co.

1971 **STEPHEN STACHENKO**
Canada & Dominion Sugar Co. Ltd.

1972 **THOMAS R. GILLETT**
California & Hawaiian Sugar Co.

1973 **WILLIAM W. BLANKENBACH**
B.C. Sugar Refining Co. Ltd.

1974 **W. AUSTIN CLARNO**
Savannah Sugar Refinery

1975 **DR. FRANK G. CARPENTER**
Cane Sugar Ref. Research Project

1976 **FRANK C. STAPLES**
SuCrest Corporation

1977 **R. STUART PATTERSON**
California & Hawaiian Sugar Co.

1978 **DR. JAMES C. P. CHEN**
Cane Tech Sugar Consultants

1978 **DR. MICHAEL C. BENNETT**
Tate & Lyle Engineering Ltd.

1979 **KENNETH R. HANSON**
Amstar Sugar Corporation

1980 **FRANKLYN M. WILLIAMS**
Calgon Corp. Activated Carbon Div.

1981 **ALBERT BERGE**
Raffinerie Tirlemontoise

1981 **WARREN L. REED**
Revere Sugar Refinery

1982 **JOHN V. LOPEZ-ONA**
The National Sugar Refinery Co.

1983 **GEORGE W. MULLER, JR.**
Executive Director, S.I.T.

1984 **JOHN B. ALEXANDER**
Tongaat-Hulett Sugar Ltd.

RECIPIENTS OF THE S.I.T. CRYSTAL AWARD**

(S.I.T. Achievement Award in Sugar Technology)

1984 **Dr. MARGARET A. CLARKE**
Sugar Processing Research Inc.

1985 **A. MARK HERTZBERG**
CSR Lmited

1985 **STANLEY GEORGE**
B. C. Sugar

1986 **DONALD K. LUKE**
Sugar y Azucar

1986 **NEIL L. PENNINGTON**
California & Hawaiian Sugar Co.

1987 **GEORGE FAWCETT**
Savannah Sugar Refinery

1987 **HEIKKI J. HONGISTO**
Finnish Sugar Company, Ltd.

1988 **ADOLPH M. BARTOLO**
Imperial Sugar Company

1989 **DR. CHUNG CHI CHOU**
Amstar Sugar Corporation

1989 **DR. R. A. MCGINNIS**
Beet Sugar Institute (Retired)

1990 **JOSEPH A. HARRISON, SR.**
Supreme Sugar Company, Inc.

1990 **EARLE J. ROBERTS**
Sugar Processing Research, Inc.

1991 **ENRIQUE R. ARIAS**
Sugar Cane Growers Coop of Florida

1992 **GEORGE C. CONRAD**
Western States Machine Co.

1993 **DAVID M. HUMM**
California & Hawaian Sugar Co.

1994 **LEON A. ANHAISER**
Imperial Holly Corporation

1995 **LUIS SAN MIGUEL BENTO**
RAR Refinarias de Acucar Reunidas

1996 **MARY AN GODSHALL**
Sugar Processing Research Institute

1997 **PETER REIN**
Tongaat-Hulett Sugar Limited

1997 **RICHARD RIFFER**
California & Hawaiian Sugar Co.

1998 **KEITH SINCLAIR**
Tate & Lyle Sugars

1999 **PETER FIELD**
CSR Sugar

2000 **MURRAY R. PLAYER**
CSR Sugar

2001 **STEPHEN J. CLARKE**
Florida Crystals Corporation

2002 **MICHAEL SASKA**
Audubon Sugar Institute, LSU

2003 **PETER G. WRIGHT**
Sugar Research Limited

2004 **MICHAEL DONOVAN**
Tate and Lyle Sugars

2005 **G. C. SINGH**
Al Khaleej Sugar Co. (L.L.C.)

2006 **DENIS BOURÉE**
Saint Louis Sucre

2006 **KLAUS E. AUSTMEYER**
Horn GmbH &Co. KG

2007 **JEAN-PAUL MERLE**
C & H Sugar Co.

2008 **JEFFREY ROBINSON**
American Sugar Refining Co.

2009 **DAVID MEADOWS**
Tongaat-Hulett Sugar Ltd.

2010 **BRIAN HARRISON**
Imperial Sugar Company

RECIPIENTS OF THE S.I.T. CRYSTAL AWARD**

(S.I.T. Achievement Award in Sugar Technology)

2011 **MIKE INKSON**
Sugar Knowledge International Ltd.

2012 **DONALD A. RYAN**
Lantic Inc. - Vancouver

2013 **AHMED VAWDA**
United Sugar Company

2014 **ROBERT FALGOUT**
Nicholls State University

2015 **PETER SIMPSON**
New Zealand Sugar C., Ltd.

2016 **CHARLEY RICHARD**
Sugar Processing Research Institute, Inc.

2017 **PETER MARAIA**
ASR Group – American Sugar Refining, Inc.

2018 **DAVID P. DEMONE**
ASR Group – American Sugar Refining, Inc.

**** Formerly – Honorary Award**

S.I.T. HONORARY MEMBERSHIP

1963 – Dan Gutleben
1964 – Edgar W. Harris
1968 – Curtis L. Taggart
1970 – George P. Meade
1998 - Earl Roberts
1999 – Bruce Foster
1999 – Mario de Campos Vidal
2002 – Marcel Braeckman
2008 – G. C. Singh
2009 – Leon A. Anhaiser
2010 - Robert Falgout
2016 - Chung Chi Chou

GROUP OF GUESTS AT THE 2018 S.I.T. BONITA SPRINGS CONFERENCE, BONITA SPRINGS, FLORIDA, U.S.A.

SUGAR INDUSTRY TECHNOLOGISTS INCORPORATED

HISTORY

Sugar Industry Technologists was founded in the United States in 1941 by several North American sugar refineries. It has since grown to an international association with a membership of about fifty corporate sugar refining companies, and about eighty corporate allied companies with over three hundred individual members from around the world.

ANNUAL TECHNICAL MEETING

The association holds an annual convention, usually in May. At this convention, technical sessions are held where members present technical papers based on research or development in cane sugar processing technology, quality control, environmental issues, and other topics of interest to the cane sugar refining industry.

Non-members may attend this conference on payment of a non-member fee equal to a membership fee.

A spousal program, social functions and an awards banquet are also held at the convention where the members have the opportunity to get acquainted and discuss mutual technical concerns.

PUBLICATIONS

Proceedings of the papers presented at the technical sessions is posted on the S.I.T. website: www.sugar industrytechnologists.com

AWARDS

Sugar Industry Technologists presents three awards and one scholarship annually.

- The Crystal Award is presented to individuals who have contributed notably to the technological advancement of the Sugar Refining Industry.
- The George and Eleanore Meade Award is presented in recognition of the best paper presented at the Annual Technical Sessions.
- The Frank Chapman Memorial Award is presented for the best poster presentation at the Annual Technical Sessions.
- The Joe Harrison Sr. Memorial Scholarship is awarded to a student attending the Cane Sugar Refiners Institute Course at Nicholls State University.

MEMBERSHIP

A Corporate membership is $450.00 USD with individual Associate membership at $40.00 USD. Affiliate memberships are available to academic and scientific persons at $60.00 USD. All fees are payable in U.S. funds. Annual fees cover the period from July 1 to June 30.

UNITED STATES SUGAR CORPORATION

SOUTHERN GARDENS CITRUS

Sugar Industry Technologists Conference

March 26, 2018

Management Report

Management Report

Report Date	3/22/2018

Safety	First aid only for insect bite

Daily Production

		Comments
Sucrose % Cane	13.2	
Milling		No major issues
Cane Ground (tons)	40107	
12-8 shift grind (tons)	13635	
Extraction	95.81	
Cane ordered today	41000	
Process		Had clarification problems on No. 2 and 6 clarifiers yesterday. This led to some bad sugar that choked the VHP dryer. Washed the dryer and got it back in service by midnight. Decanter 3 broke the shear pin yesterday and choked with mud. Working today to get it back in service.
Yield lb / ton cane	222	
Pol % Filter Cake	3.5	
Pol % Decanter mud	3.0	
Final Molasses purity	33.7	
Final Mol % cane	3.3	
Refinery		Struggled with filtration. Did all PM's on pans yesterday while we were slowed. It has improved this morning but not enough to build level in the ILS tank.
Refinery Melt	47476	
Refinery Production	46460	
Power Plant		Fuel used for burner PM's. No issues today
Steam / ton cane	811	
Fuel oil to boilers	214	

Time Account — *insert reasons for lost grinding totaling greater than 500 tons*

	Tons lost	*Reason*
Ag		
Rail Road		
Milling	1012	Stopped C tandem to clean tracks
Planned Main		
Power Gen		
Boiling House		
Cane Supply Rain		
Other	47	

UNITED STATES SUGAR CORPORATION

SOUTHERN GARDENS

Daily Packaging

Clewiston Refinery
Daily Packaging / Loading & Shipping Report

Ref. Op. Report:	132									Date: March 12, 2018
Operating Days-M	12									Julian Date:
		PRODUCTION				SHIPMENTS				
		CWD 3/12/2018	CWD March	USSC Yr to Date	USC Yr to Date	CWD 3/12/2018 Shipped	CWD March	USSC Yr to Date	USC Yr to Date	Sales Returns Total for Mo.
Packaging:										
Big Sig		3,540.40	32,026.00	421,132.80	594,959.40	432.00	62,882.80	530,774.40	784,916.00	-
Little Sig	2# Line	-	-	77,052.60	86,201.80	-	1,989.60	76,867.20	89,038.80	-
Little Sig	4# Line	2,792.40	30,835.60	118,104.40	197,797.00	-	-	-	-	-
Little Sig	5# Line	-	-	36,481.60	46,615.60	-	96.00	41,016.00	46,512.00	-
		-	-	-	-	-	-	-	-	-
50# Fine Gran - Line 1		8,675.00	67,825.00	767,075.00	1,017,100.00	22,850.00	131,125.00	1,699,545.00	2,324,192.00	425.00
50# Bakers - Line 1		-	-	-	-	450.00	2,350.00	31,925.00	42,000.00	-
50# Fine Gran - Line 2		5,475.00	84,125.00	962,050.00	1,348,375.00	-	-	-	15,975.00	-
50# Bakers - Line 2		-	3,252.90	33,535.90	43,821.90	-	-	-	-	-
25# Fine Gran - Line 1		-	14,058.25	68,472.75		24.50	16,977.50	69,602.50	125,266.50	-
25# Fine Gran - Line 2		-	-	-	-	-	-	-	-	-
Packaging		**20,482.80**	**222,122.75**	**2,473,905.05**	**3,334,870.70**	**23,756.50**	**215,420.90**	**2,449,730.10**	**3,427,900.30**	**425.00**
Bulk Loading:	# Cars									
Rail:	10	21,923.200	258,705.600	3,081,615.200	4,284,108.700	23,954.500	252,139.300	3,077,444.000	4,297,230.200	
Remelt	0	-	863.000	13,982.000	18,431.040	-	-	-	-	
Bulk Trucks	15	7,490.800	67,419.000	660,993.300	943,314.100	7,490.800	67,419.000	660,993.300	943,497.100	
Sales Return		-	-	-	-					-
Bulk		**29,414.000**	**326,987.500**	**3,756,590.500**	**5,245,853.840**	**31,445.300**	**319,558.300**	**3,738,437.300**	**5,240,727.300**	
Total Granulated		**49,896.800**	**549,110.250**	**6,230,495.550**	**8,580,724.540**	**55,201.800**	**534,979.200**	**6,188,167.400**	**8,668,627.600**	**425.00**
Liquid Sugar:										
LS Trucks	9	3,199.230	27,853.200	346,539.602	474,040.960	3,199.230	27,853.200	346,178.073	473,679.432	-
Total Granulated & Liquid:		**53,096.030**	**576,963.450**	**6,577,035.152**	**9,054,765.500**	**58,401.030**	**562,832.400**	**6,534,345.473**	**9,142,307.032**	**425.00**
Delta Inventory:										
Beginning Silo Inventory		71,911.00								
Ending Silo Inventory		74,212.00								
Delta Silo Inventory		**2,301.00**								

		Accum. Month	USSC YTD	USC YTD	Whse. Inv.	Cars not shipped
DAILY GROSS TOTAL PRODUCTION:	55,397.030	607,765.460	6,590,143.152	9,184,423.750	56,230.45	5,759.40
DAILY NET TOTAL PRODUCTION:	55,397.030	606,902.460		9,165,992.710		

3

Summary of Cane Quality Performance Factors

Summary of Cane Quality Performance Factors

Criteria		Date: 03/21/2018	Values		
Cane in System (Carloads)		Static Target =	0-350	Dynamic Target =	-
Time Out of Spec (hr)		Static Target =	3	Dynamic Target =	
	Average to Date		6.34		
	Average Previous 7 Days		5.29		
	Yesterday		12.50		0.00
Bank 8 AM (Carloads)		Static Target <	120.00	Dynamic Target <	
	Average to Date		141.04		
	Average Previous 7 Days		148.29		
	Yesterday		159.00		159.00
	Today		162.00		162.00
Cane Age (Hrs)		Static Target <	8.00	Dynamic Target <	
	Average to Date		7.66		
	Average Previous 7 Days		7.28		
	Yesterday		7.35		7.35
Old Cane (% of Total)		Static Target <	7.50	Dynamic Target <	
	Average to Date		3.83		
	Average Previous 7 Days		0.71		
	Yesterday		0.10		0.10
Dextran Levels					
Crusher Juice (ppm on Brix)		Static Target <	1,000.00	Dynamic Target <	
	Average to Date		1,070		
	Average Previous 7 Days		0		
	Yesterday		0		
Raw Sugar (MAU)		Static Target <	100.00	Dynamic Target <	
	Average to Date		68		
	Average Previous 7 Days		0		
	Yesterday		0		
Factory Milling Performance					
Cane Ground / Requests (%)		Static Target =	100.00	Dynamic Target =	
	Average to Date		90.75		
	Average Previous 7 Days		99.24		
	Yesterday		100.00		100.00
RR Performance					
Average Load Station Wait Time (hr)		Static Target <=	2.50	Dynamic Target <=	
	Average to Date		2.26		
	Average Previous 7 Days		2.35		
	Yesterday		2.43		2.43
Long Ld. Station Waits (> 4hr)(% of Total)		Static Target <	10.00	Dynamic Target <	
	Average to Date		8.08		
	Average Previous 7 Days		9.68		
	Yesterday		13.26		13.26
Long BRY Yard Waits (>2hr)(% of Total)		Static Target <=	10.00	Dynamic Target <=	
	Average to Date		1.56		
	Average Previous 7 Days		2.64		
	Yesterday		18.45		18.45
Average Factory Yard Wait (hr)		Static Target <=	3.50	Dynamic Target <=	
	Average to Date		3.24		
	Average Previous 7 Days		2.85		
	Yesterday		2.80		2.80
Long Yard Waits(>4 hr)(% of Total)		Static Target <	30.00	Dynamic Target <	
	Average to Date		27.76		
	Average Previous 7 Days		19.16		
	Yesterday		13.00		13.00

Printed: 03/22/2018 Report: rptScoreCard

Mill Total Losses (Tons)
Crop 2016/17

Mill Total Losses - Crop 2016/17

Cane Supply	**431,263**
Milling	**343,186**
Power Generation	**134,621**
Planned Maintenance	**99,040**
Boiling House	**86,819**
Start-Up	**52,202**
Grand Total	**1,147,131**

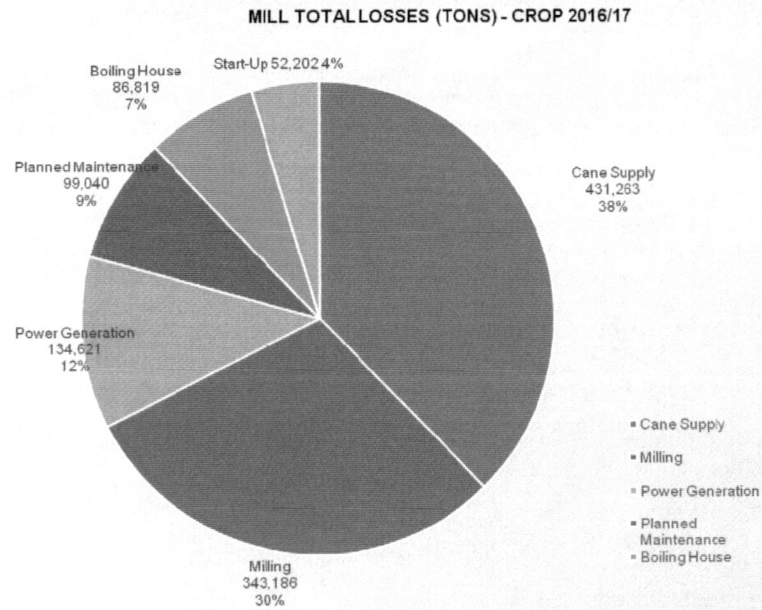

Customer Complaint Rate
Complaints per 100K Shipped

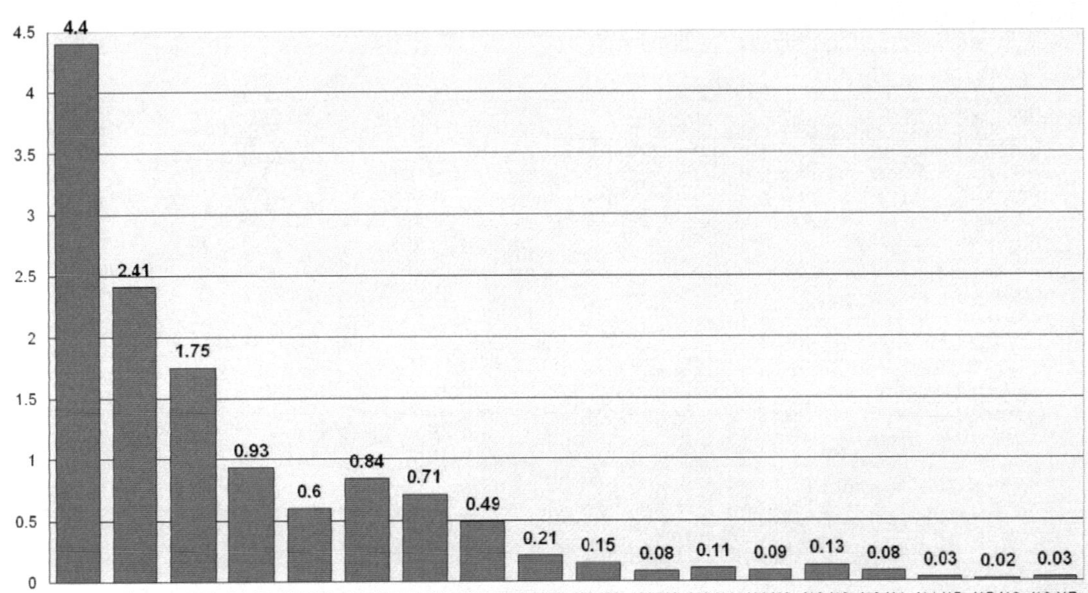

Mill Total Losses (Tons)
Crop 2016/17

Mill Total Losses - Crop 2016/17

Cane Supply	**431,263**
Milling	**343,186**
Power Generation	**134,621**
Planned Maintenance	**99,040**
Boiling House	**86,819**
Start-Up	**52,202**
Grand Total	**1,147,131**

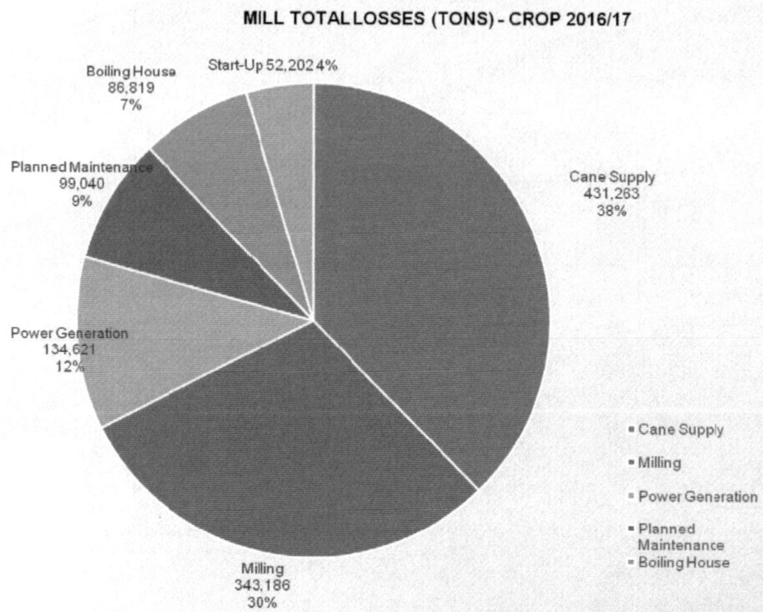

UNITED STATES SUGAR CORPORATION SOUTHERN GARDENS

7

Production

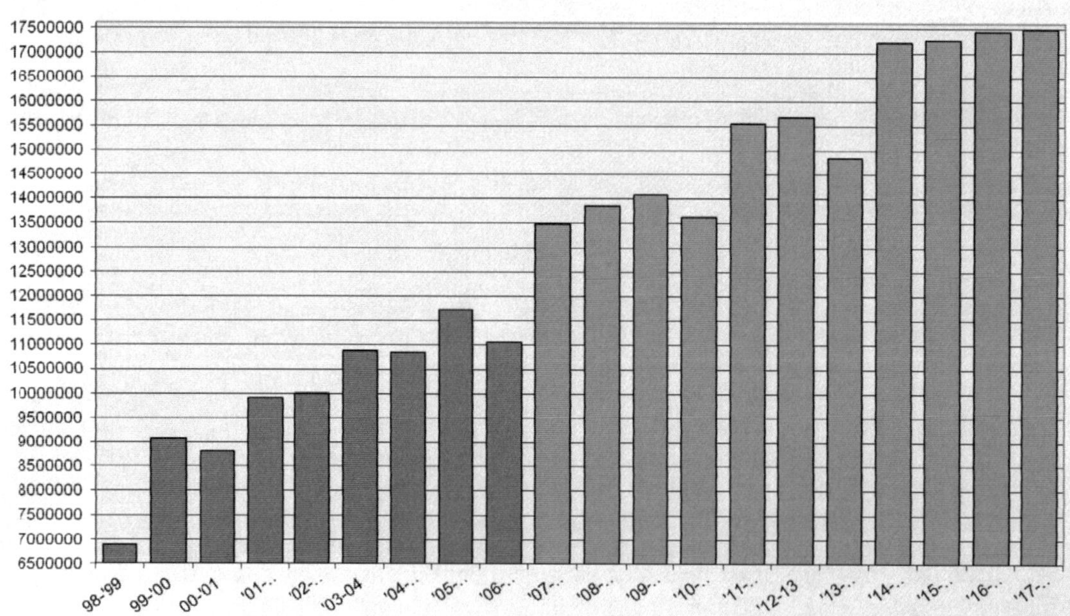

UNITED STATES SUGAR CORPORATION SOUTHERN GARDENS

8

Head Count

Pre-initiative: 11,999 per cwt

Post-initiative: 39,724 per cwt

9

UNITED STATES SUGAR CORPORATION

SOUTHERN GARDENS

#1181 - Developments in the repeatability of Cooling Crystalliser performance

Chris Mayhew, Carlton Haynes, Robert Howe, David Cooper and Craig Parker, British Sugar

Several alternatives exist for introducing seed crystals in vacuum pans (e.g. Ditmar seeding, Mother/daughter systems and Brunswick seeding). British Sugar predominantly uses Cooling Crystalliser technology to generate the Brunswick Seed for its white pans. As reported at the 2017 SIT conference in Taiwan (paper #1153 – Linear Pan Growth)[1], the operational consistency of a batch white pan and the resultant crystal particle size of the final product is very closely related to the consistency of the seed crystals introduced (particle size and number).

This paper reports on the developments at both the Bury and Wissington factories between 2016 and 2018, focussing on improving the repeatability of the Brunswick Seed produced. It will also present British Sugar's current approach to the operation of Cooling Crystallisers and our developments towards what we see as best practice.

Introduction

During the 2016-17 beet sugar campaign at British Sugar's Wissington factory a project, working on developing a Linear Pan Growth (LPG) technique to improve white crystal quality, identified issues with the production of the seed used in the white pan[1]. These issues concerned the quantity, quality and repeatability of seed produced by the Cooling Crystalliser system and these were identified as critical to the success of the LPG project. This paper reports on the developments that have taken place to improve the seed quality during the 2017 Juice Run and 2017-18 beet sugar campaign.

Background

The production of quality white sugar from white massecuite depends on a reliable source of seed crystal, in terms of quantity, quality and repeatability. A common source of seed material used within the sugar industry is milled slurry; typically granulated sugar that is ground down, in either a ball or pin mill, to a particle size of around 4-10μm. These particles are milled in a carrier fluid, such as dry Iso-Propyl Alcohol (IPA), which does not dissolve the sugar but keeps the particles from clumping together before they are added to the batch pan. IPA is often chosen as it evaporates when it comes into contact with hot juice, leaving the seed particles behind to grow in the juice. Glycerol is sometimes added to the slurry to further reduce the potential for crystals to clump.

The main challenge, with using this slurry to seed the batch pans, is to ensure only the seed particles added grow to produce the final product. In reality, this controlled growth is threatened by the early conditions in the batch pan. The fragility of the 4-10μm particles is highly dependent of controlling the mother liquor Degree of Saturation (D_{SAT}) within the range of 1.0 to 1.15 (Metastable zone). Running below this range results in the seed dissolving away, whereas operating above the 1.15 range may result in additional seed sites (false grain) spontaneously being produced. In either case, the number of original seed sites presented to the pan is subsequently altered. This reduces the overall control and repeatability of the final crystal product quality.

During the massecuite's passage through the calandria, it is exposed to the hot tube wall temperature, which raises the local massecuite temperature sufficiently to reduce the mother liquor saturation below the minimum 1.0. At this point some of the small seed particles may dissolve completely, reducing the overall crystal population at the start of the batch cycle. At the same time, where the massecuite meets the vapour space, the operating vacuum will result in a lower massecuite temperature at this interface. This will increase the mother liquor D_{SAT}, potentially risking the spontaneous formation of false grain if the D_{SAT} rises above the 1.15 upper level of the metastable zone.

The potential risk of both these conditions affecting the quantity of seed sites, present in the initial phase of the pan cycle, can be reduced by increasing the size of the seed particles added to the pan. Larger crystals will be less likely to dissolve completely when passing through the calandria. They also provide a greater surface area for crystallisation when at the massecuite/vapour interface, reducing the potential for false grain to form. Therefore, introducing a larger seed for the batch pans offers greater final crystal quality control potential.

White pan seeding options

There are several options available to the sugar technologist to produce a larger seed size for batch pans, including; mother – daughter pans, recovery sugar magma and Cooling Crystallisation.

Mother – daughter pans involve growing seed in one "mother" batch pan and then sharing the massecuite of this pan as the seed footing for "daughter" batch pans. This mechanism still requires the mother pan to be seeded using a slurry system and therefore is still vulnerable to the associated issues discussed earlier.

Recovery sugar magma uses the sugar produced from centrifuging massecuite from a later boiling stage within the refinery. This method involves mixing this sugar with saturated syrup, before using the resultant magma to seed the batch pans. Here the control of the seed size is dependant of the performance of the later boiling stages. Any significant variability will translate back to the recipient pan and ultimately into the final crystal particle size. Colour inclusion may also cause problems, as colour formed in the later boiling stages can become trapped within the seed, subsequently getting locked into the recipient sugar product.

Cooling crystallisation grows the seed slurry from its initial 4-10μm size up to a desired particle size, by controlling the D_{SAT} of the mother liquor through temperature control of the massecuite. This method takes a concentrated syrup charge and cools it until the D_{SAT} of the syrup reaches the target saturation for seeding, when the seed slurry is then added. This seed is then grown on through the controlled cooling of the massecuite to maintain the mother liquor at the optimum D_{SAT} profile.

Evaporation vs Cooling Crystallisation to drive saturation

Whichever seeding mechanism is selected, the first step is to raise and control the syrup concentration to elevate the sucrose concentration above its saturation point. The D_{SAT} of the syrup influences how quickly the crystals grow; the higher the D_{SAT}, the faster the growth rate. However, increasing the D_{SAT} too high will lead to additional (false) grain spontaneously forming and a subsequent loss in quality control of the final crystal product. The finer art in controlling crystal quality during the growth phase is critically linked to D_{SAT} control.

There are two mechanisms readily available to the sugar technologist to control the mother liquor D_{SAT}; evaporation and cooling.

- ***Evaporation*** concentrates the mother liquor by driving off water, increasing the mother liquor brix and hence raising its D_{SAT}. Evaporation requires a source of heat to convert some of the water in the mother liquor into vapour. This heat is typically supplied by condensing steam in the batch pan's calandria and transferred into the massecuite via the surface of the calandria. This mechanism raises the temperature of the massecuite as it passes through the calandria, reducing the mother liquor's D_{SAT} below 1.0 next to the calandria's surface. Once through the calandria, the heat in this massecuite stream then dissipates through the rest of the massecuite as it mixes in the pan. This heat eventually drives water evaporation from the mother liquor as it circulates towards the upper layer of the massecuite, increasing the mother liquor brix (and hence D_{SAT}).

- ***Cooling*** the massecuite lowers the solubility of the sucrose in the mother liquor and hence increases its D_{SAT}, driving crystal growth. However, this mechanism is limited by the water that is left in the mother liquor, by the physical constraints with viscosity and reduced sucrose diffusion rate through the cooler mother liquor. Heat transfer from the massecuite is delivered by the use of cooling coils around the periphery of the Cooling Crystalliser vessel. This relies on good control of the cooling water temperature and an even flow of massecuite over the cooling coils to prevent over cooling at the coil surface. Poor cooling control may result in localised cold spots, with the risk of elevated D_{SAT} and spontaneous false grain formation.

Both methods have their benefits, but also come with issues. In producing a consistent seed material for batch pans, evaporation (using the mother daughter system) can provide a quicker route, with potential to produce a larger grain size. However, this method involves controlling mother liquor concentration by a combination of temperature (both steam and vacuum profiles) and feed syrup, which may also vary in concentration, purity and temperature over the batch cycle. At a typical feed syrup purity of 95, a 1 brix change in mother liquor concentration effects the D_{SAT} by around 0.07, taking a pan controlling at 1.12 D_{SAT} into a potential false grain range of 1.19 D_{SAT}.

Cooling crystallisation is typically slower, with growth rates diminishing as the massecuite temperatures fall. The method is also limited in terms of crystal content as the water content of the massecuite remains unchanged over the cycle of the batch. However, with a single control variable during the growth phase (massecuite temperature), optimisation of the growing profile is simpler and more repeatable. In comparison with the evaporation example above, a 1°C decrease in pan temperature will increase the D_{SAT} by only 0.013.

Cooling Crystallisation plant

At all four British Sugar beet factories, the seed for batch vacuum pans is based around a milled sugar slurry. This slurry is a mixture of granulated sugar and IPA that is milled for a set time in a ball mill. At British Sugar's Bury, Cantley and Wissington factories, the milled slurry produced is grown from 4-7μm to a more robust 150-160μm, via Cooling Crystallisation, which is detailed below in figure 1

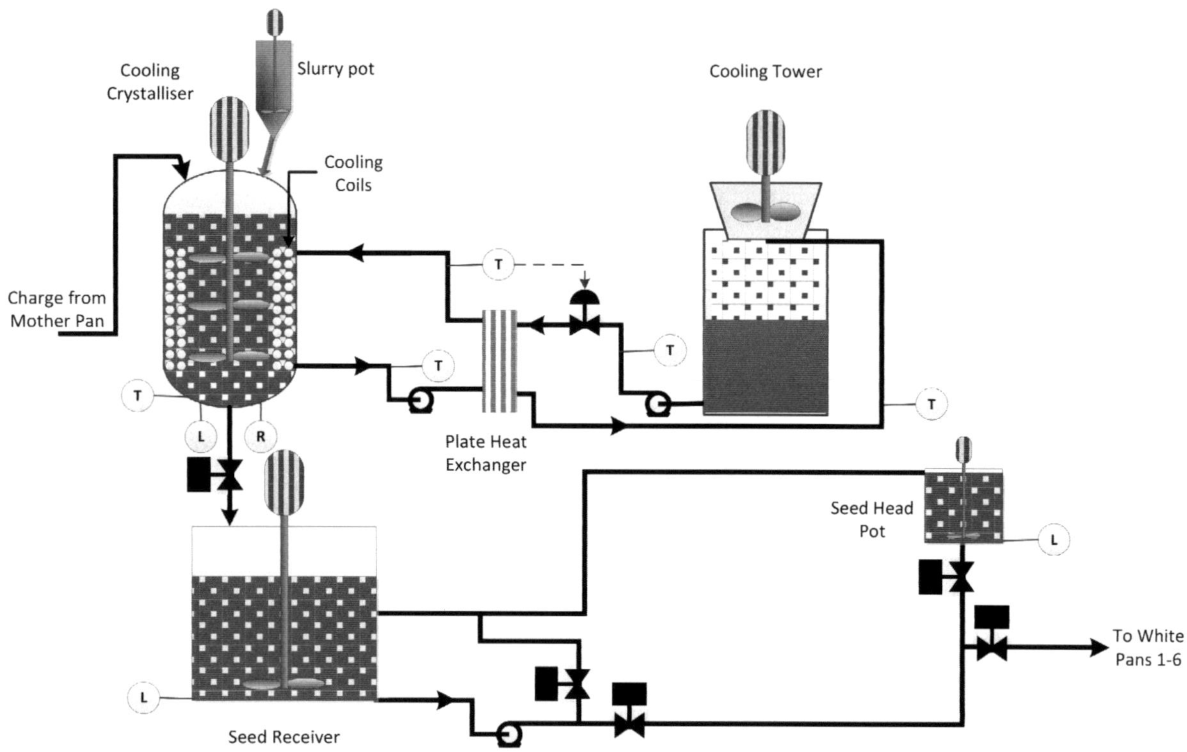

Figure 1 A typical Cooling Crystalliser installation

- A charge volume for the Cooling Crystalliser is prepared in a batch pan from a measured volume of standard liquor. This charge is then boiled down to meet the target transfer brix.
- The charge is then pumped into the Cooling Crystalliser vessel, the stirrer started and the juice is then cooled to the seed temperature (set to achieve 1.14 D_{SAT}).
- When the contents of the Cooling Crystalliser meet the target seed temperature, a measured volume of seed slurry is added to the vessel. This seed slurry disperses in the cooled juice and starts to grow.
- The juice (now massecuite) is then cooled along a controlled temperature profile, to maintain a desired crystal growth rate by controlling the D_{SAT} of the mother liquor.
- Once the massecuite has reached its temperature end point, cooling is stopped and the crystalliser contents transferred to the seed receiver.
- The white pans are seeded with massecuite from this seed receiver.

Reliable repeatable seed quality

The decision to install the Cooling Crystalliser seed systems was based on providing a consistent quality seed for the batch pans, especially the white pans. These systems, installed during the 1980's, were part of the pan floor automation activities, which aimed to produce a robust and repeatable seed for the factories. However, during the 2016/17 beet campaign, a project to develop a LPG approach for the white pans, identified that the seed quality from the Cooling Crystalliser system was variable, and to a large extent uncontrolled in terms of the crystal size. It became evident that the variability of the seed massecuite was adversely impacting on the LPG project, as well as affecting the factory's white sugar production performance indicators[1].

This finding led the authors to study the Cooling Crystalliser system at British Sugar's Wissington factory with the aim of developing improved understanding of how to optimise the repeatability of Cooling Crystalliser systems.

Factors affecting Cooling Crystalliser performance

- ***Transfer brix***

One of the key factors separating batch pan seed production from Cooling Crystalliser systems is the charge condition. With a batch pan, the pan charge is boiled down, seeded and then fed with standard liquor. In the case of the Cooling Crystalliser, once it is charged with concentrated liquor no more syrup is added at any point in the crystal growth cycle. To maintain a consistent final crystal content from the Cooling Crystalliser, it is critical to start with a fixed charge volume and syrup brix. Changing either of these factors will result in a different mass of sugar in the Cooling Crystalliser, which will alter the final seed crystal size and subsequently lead to batch pan crystal quality issues.

At Wissington, the Cooling Crystalliser is charged with syrup that has been concentrated by a dedicated batch pan, with the end point of the pan set to meet a target transfer brix. The contents of the pan are then pumped into the Cooling Crystalliser, the brix checked and diluted, if necessary, to meet the target start conditions for the cooling profile. For the LPG project, the Cooling Crystalliser was fitted with a K-Patents refractometer, to help study the mother liquor conditions in the vessel during the cooling phase.

Once commissioned and calibrated, the refractometer's variance was measured at less than 0.2 brix against laboratory analysis. This accuracy highlighted the variability in the charge liquor brix between batches leaving the concentrator pan. It was then decided to introduce a dilution phase in the Cooling Crystalliser program and use the refractometer's output to control the dilution of the charge down to a target seed brix. This proved extremely successful in removing the variability of the seeding brix.

- ***Seeding temperature vs. liquor brix***

With a steady charge liquor purity and consistent seeding brix, the seeding temperature was set to achieve the target seeding D_{SAT} of 1.14. If the transfer brix was above target, the system would dilute down to the target seeding brix with water and then cool to the seed temperature. However, due to the fixed target seeding temperature, if the transfer brix was below the seeding target, the program would add the seed too early. In the worst case scenario this could result in the seed dissolving completely, with a negative impact to refinery operations (quality and productivity). To protect against this scenario, a step was added to the Cooling Crystalliser program to adjust the target seeding point temperature to meet the 1.14 D_{SAT} target. This was based on the brix reading from the refractometer in the Cooling Crystalliser. A lower limit on the transfer brix range ensured that the system did not attempt to seed the crystalliser if the transfer brix was too low.

- ***Cooling profile***

Once charged (brix adjusted if required), the system is cooled to raise the D_{SAT} of the mother liquor up to the desired 1.14. The cooling element within the Cooling Crystalliser is a double bank of cooling coils around the internal periphery of the Cooling Crystalliser vessel. Cold water is pumped through the coils, with the flow rate and inlet temperature controlled. This is to minimise the risk of false grain forming on the tube walls from over saturating the liquor adjacent to the tubes.

After the crystalliser is seeded, the massecuite is cooled along a temperature profile (driving the saturation profile) to maintain a desired growth of the crystals over the cycle. The cooling continues until a final end temperature is achieved within the massecuite. At this point the cooling is turned off and the massecuite remains, circulating in the vessel, until there is sufficient space in the seed receiver to accommodate the contents of the Cooling Crystalliser vessel. The original cooling profile can be seen in figure 2.

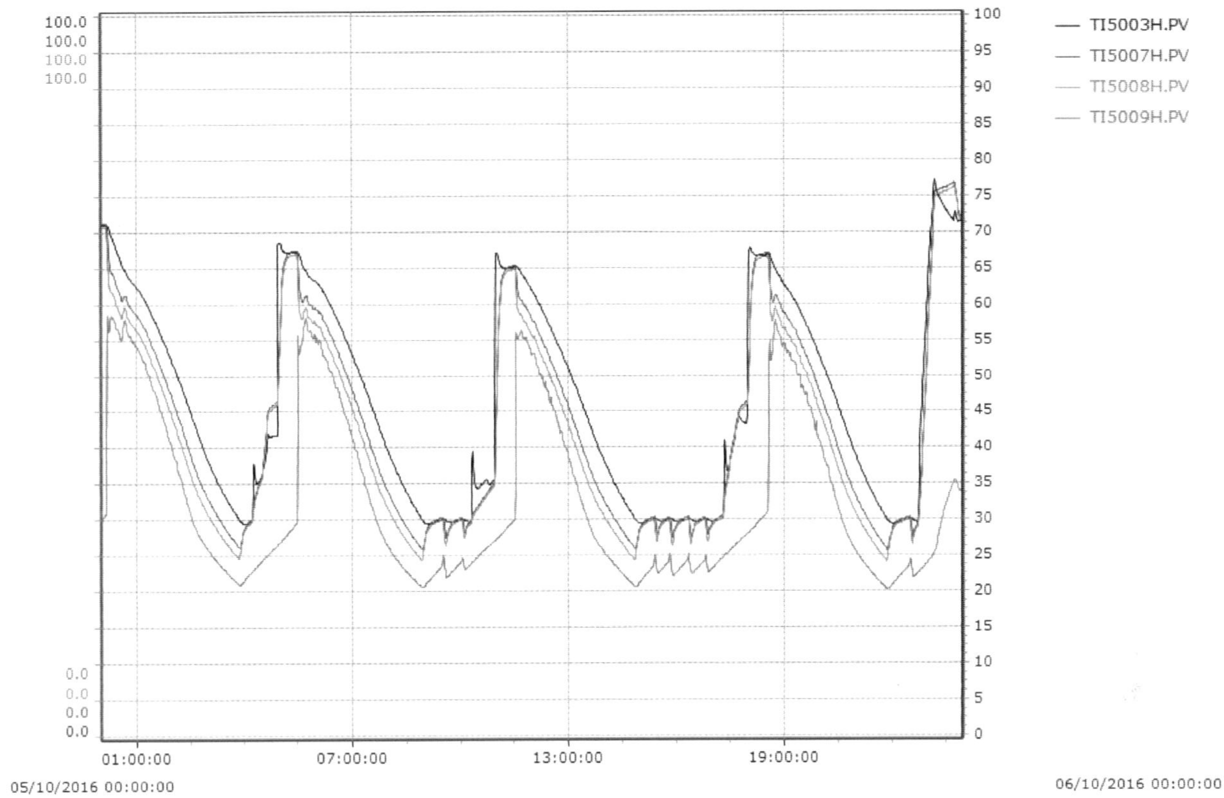

Figure 2 Original cooling profile

- **TI5003H.PV – massecuite temperature,**
- **TI5007H.PV – cooling water temperature leaving the cooling coils**
- **TI5008H.PV – cooling water temperature entering the cooling coils,**
- **TI5009H.PV – cooling water temp from cooling tower to plate heat exchanger**

Typically, under the original cooling profile, the crystalliser would produce between 4 and 5 batches per day (equivalent to 80-100m^3 of seed massecuite per day).

With the output from the Cooling Crystalliser refractometer, it was possible to monitor the reduction in the mother liquor brix and hence calculate the crystal growth within the massecuite (this is covered in more detail later in this paper). This information was used to adjust the cooling profile to increase the crystal content from each crystalliser cycle. This increased crystal content was then used to release extra capacity from the Cooling Crystalliser. The modified cooling profiles can be seen in figures 3 and 4.

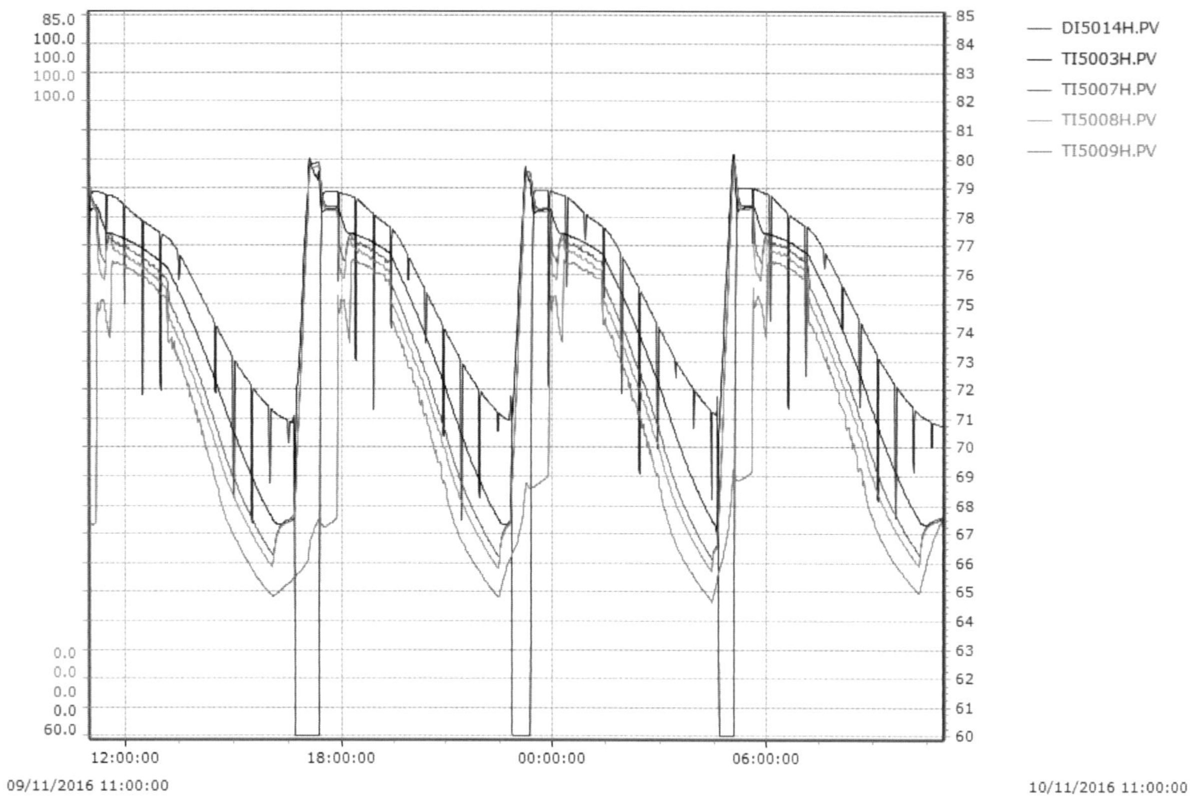

Figure 3 Original Cooling Crystalliser profile with refractometer data

When the Cooling Crystalliser refractometer was switched on, the first profiles (shown in figure 3) indicated how the transfer brix varied between cycles. It also showed how the mother liquor brix decayed during the rapid cool phase, prior to seeding. Detailed inspection of this phase of the cycle identified the formation of false grain occurring before seeding, possibly due to the over saturation of the liquor close to the cooling coils [Note: the sharp spikes on the refractometer trend are caused by the lens flushing water, used to keep the prisms clean].

With the ability to monitor the mother liquor brix, the rapid cooling phase was modified prior to seeding to reduce the ΔT cooling impact on the un-seeded liquor in contact with the cooling coils. The resultant profile (see figure 4) removed the large ΔT between the cooling coil temperature and the bulk liquor temperature during the pre-seed fast cool phase. This change reduced the rate of fall in the liquor brix, prior to seeding. At this point in the project, although vastly improved, the fall was not completely eradicated, with a drop of 0.1°Brix still evident.

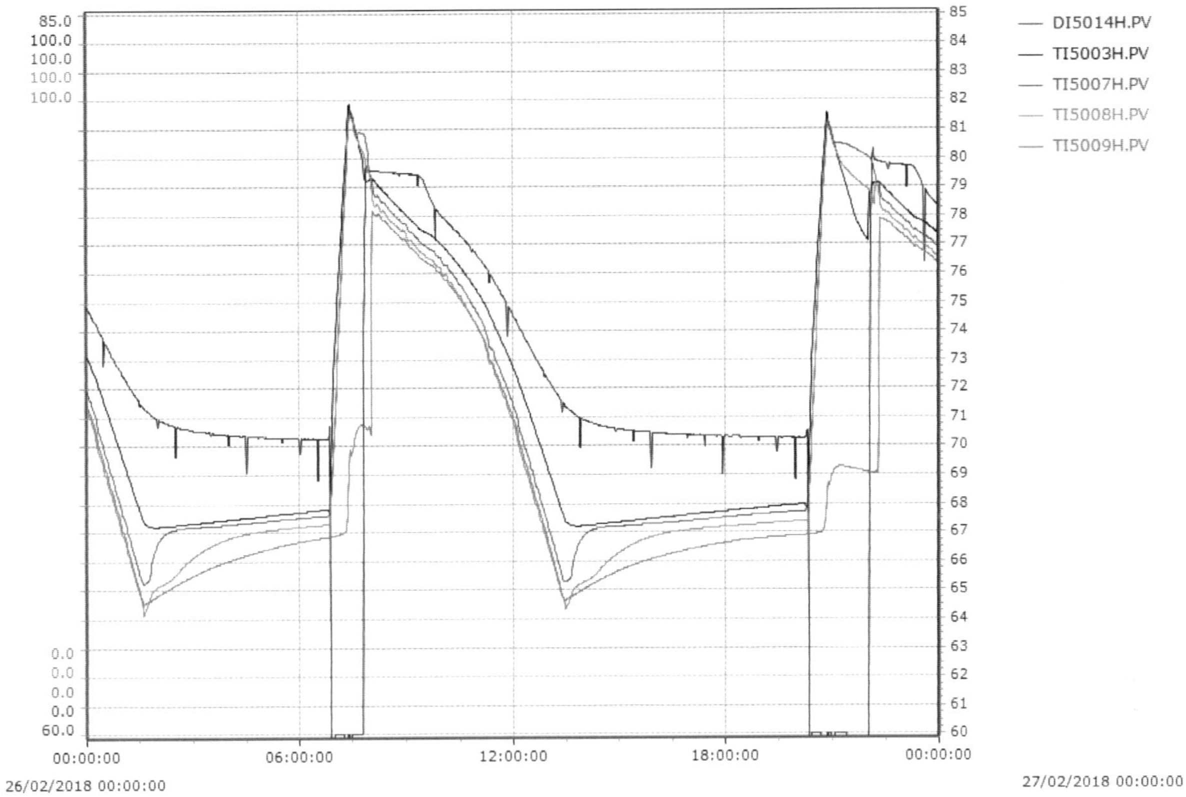

Figure 4 Cooling Crystalliser profile with revised cooling mode

- ***Seed addition***

Seed slurry is added to the Cooling Crystalliser from a stirred seed pot, fitted to the top of the Cooling Crystalliser vessel. The seed pot, which is continuously stirred when in use, is filled manually by the operator and automatically discharged into the Cooling Crystalliser when the liquor temperature reaches the target seed point. It is then incorporated into the liquor via the mixing action from the crystalliser's stirrer.

Under normal operating conditions, the Cooling Crystalliser is seeded within the temperature range of 70-73°C. The IPA rapidly evaporates, leaving the milled sugar crystals behind in the liquor. This seed delivery system requires good mixing of the slurry into the liquor, to ensure that the individual seed crystals are evenly dispersed. During the early stages of the project it was observed that very shortly after seeding, rafts of crystals were regularly visible in the crystalliser's microscope. These rafts could also be seen as fine white flakes in an observation sight glass, fitted to the side of the crystalliser's wall.

A visual inspection of the liquor surface in the Cooling Crystalliser (through the top access point), showed the surface layer circulating around the stirrer shaft, with a few small eddy vortices slowly orbiting around the shaft. The seed slurry was added to a point outside of this stirred zone and remained on the surface until it was drawn into one of the vortices and then dragged down into the bulk liquor. Typically, portions of the slurry could remain on the surface for up to 90 seconds, during which time the IPA would completely evaporate. This left seed crystals behind on the surface in clumps, before they were eventually drawn down into the bulk liquor. These surface clumps could explain the formation of the seed rafts, as seen in the microscope (Fig 5).

Figure 5 Slurry clumping on surface and a seed raft

- ***Seed inlet development comparison with the Bury St Edmunds Cooling Crystalliser***

Following the initial observations concerning the impact of the seed inlet position, discussions with colleagues at the Bury St Edmunds factory, confirmed they had experienced similar issues with seed addition and inclusion into the bulk liquor. They also reported seeing seed rafts on the crystalliser's microscope and realigned their seed injection towards the stirrer shaft.

Following some cross site discussions, the Bury team modified the seed addition rate, by pulsing the seed inlet valve, to reduce the volume of slurry accumulation on the surface during the seeding step. The team reported that both these changes reduced the number of seed rafts observed on the crystalliser's microscope.

Figure 6 Bury Seed addition adjusted to central point

- ***Relocation of the seed addition point***

In an attempt to improve the incorporation of the seed slurry during the seeding phase, the seed addition point was relocated to direct the flow of the slurry closer to the stirrer shaft. At the same time, the flowrate of the slurry was reduced to minimise the volume of slurry that would remain on the surface in clumps. Both modifications improved the incorporation of the seed slurry into the liquor, reducing the dwell time on the liquor surface down to 5-10 seconds. However, this time was still sufficient for a significant proportion of the IPA to evaporate, allowing some of the seed to clump before being drawn into the bulk liquor.

- ***Modifications to the Stirrer configuration***

It became evident that the lack of surface movement was a limiting factor in drawing the seed slurry down into the bulk liquor, before the IPA had evaporated. At this point we consulted with the stirrer manufacturers (SPXflow). We enquired whether the addition of a supplementary set of blades, near the liquor surface, would improve the surface agitation. SPXflow undertook some CFD modelling of the fluid movement within the vessel. This confirmed the original observations and indicated a potential performance enhancement from an additional blade set was achievable. The stirrer blades were specifically designed to draw down more fluid from the surface zone, whilst ensuring the liquid shear rate at the stirrer tips did not create the potential for false grain formation. The additional set of blades were manufactured by SPXflow and fitted prior to the start of the 2017/18 beet campaign. Figure 6 shows part of the additional blade set at the top of the stirrer shaft.

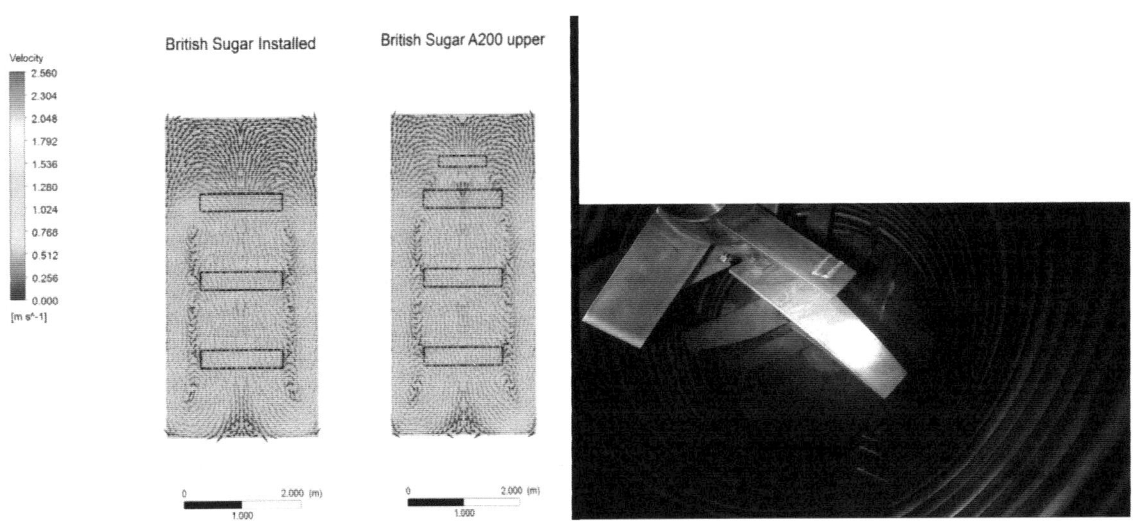

Figure 7 Stirrer Modifications CFD (left) and top set as installed (right)

After fitting the new stirrer blades, the modelling prediction was confirmed with a far greater degree of agitation seen on the surface, culminating in a larger vortex around the stirrer shaft. A further, small modification to the slurry inlet position ensured that the seed was added directly into the vortex. The visible result was the virtually instantaneous draw down of the seed slurry into the bulk of the liquor, and subsequently improved dispersion of the seed crystal into the bulk liquor.

Figure 8 shows a timeline of changes to the Cooling Crystalliser seed addition and their impact on the quality and regularity of the seed produced.

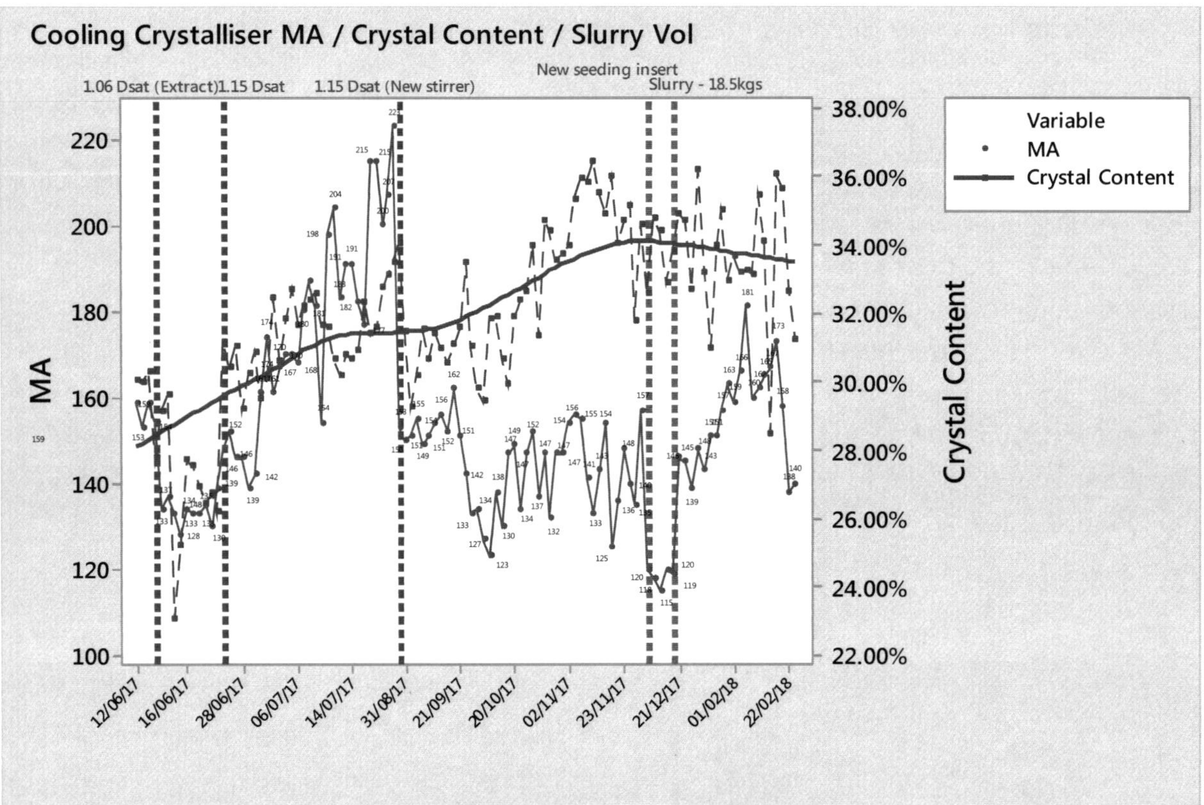

Figure 8 Impact of changes on Cooling Crystalliser performance (seed quality)

From figure 8, the most significant change appears to be the installation of the additional stirrer blades, which reduced the MA (blue line) from around 200 to 140μm.

Using the refractometer brix data

Once installed and commissioned, the refractometer output became the focus of the cooling profile development. Initially, driven by the modelling work from Chris Rhoten[2], the cooling profiles were changed to deliver a more gentle approach to the early growth phase in the Cooling Crystalliser (see figure 9). The cooling then became more aggressive towards the later stages of the cycle. When combined with the earlier changes in the seed inlet position, the crystal content of the crystalliser's massecuite rose from the original 22% to 32% crystal content.

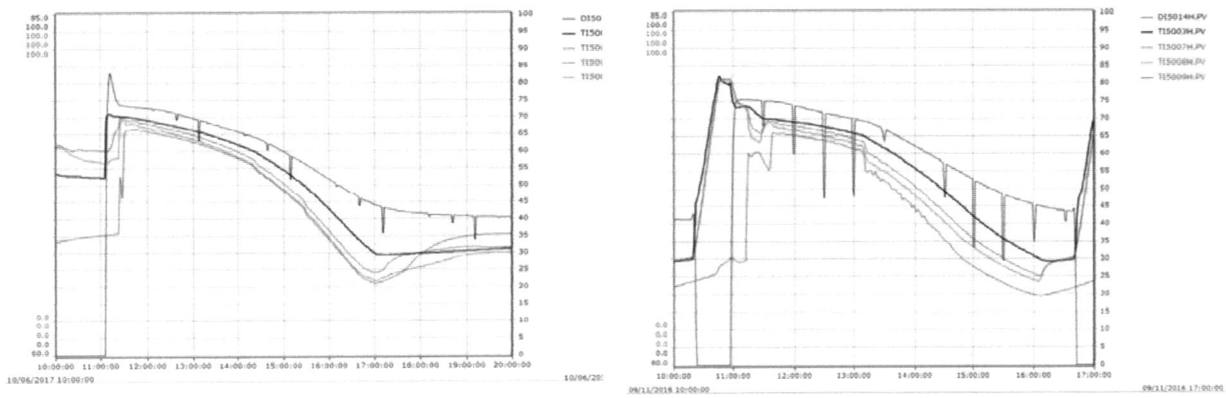

Figure 9 Original Cooling profile (left) and revised cooling profile (right)

- ***Calculating an online D_{SAT}***

During the 2017/18 campaign, the brix output from the refractometer was used in conjunction with the massecuite temperature, to calculate the D_{SAT} of the mother liquor in the Cooling Crystalliser during the cooling profile. This method used the syrup purity, combined with the initial diluted liquor brix (pre-seeding), to calculate the liquor's D_{SAT} up to the seed point. As no further dilution takes place in the Cooling Crystalliser during the crystal growth phase, this seed point brix and purity are also representative of the massecuite brix and purity throughout the cooling process.

After seeding, the refractometer output now solely represents the brix of the mother liquor. Therefore, using both the pre-seed and post seed refractometer brix values, it is possible to calculate the increase in crystal content of the massecuite and subsequently calculate the mother liquor purity (by a combination of mass and sugar balances).

Armed with this information, the massecuite temperature and measured solubility coefficients, the current D_{SAT} of the mother liquor was calculated. This information was added to the control system trends, allowing the operators to track any changes in the D_{SAT} during the cooling profile. An example of the early profile can be seen in figure 10.

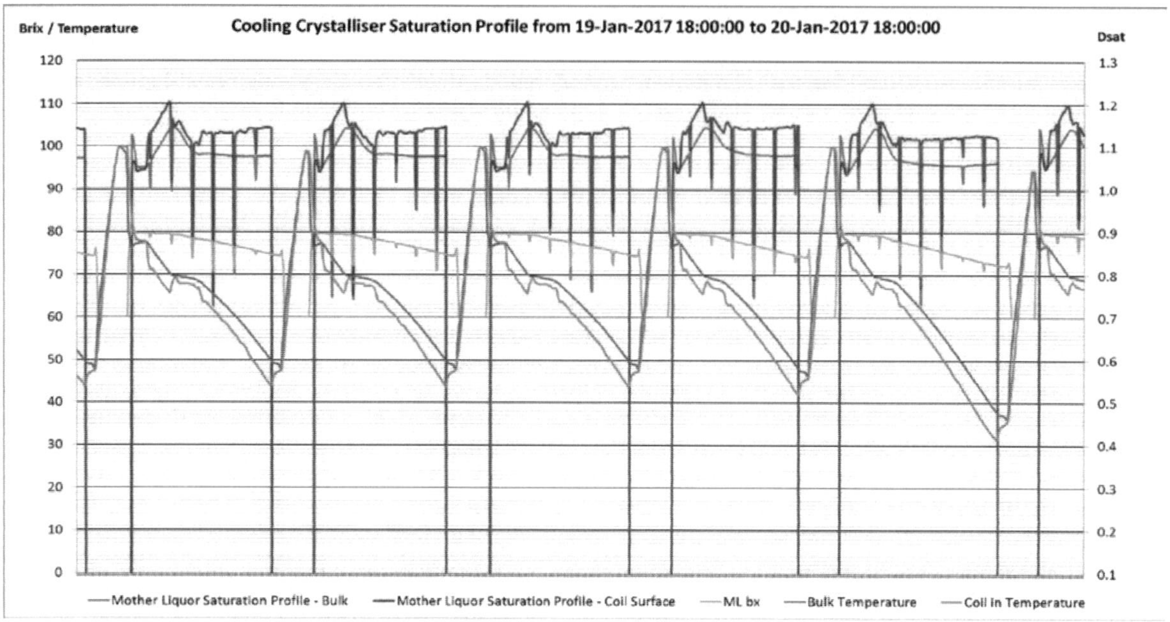

Figure 10 Original Cooling Crystalliser profile with calculated D_{SAT} values at the coils (red line) and in the bulk liquor (blue line)

When operating Cooling Crystallisers, is to essential to minimise the risk of over saturating the massecuite in contact with the cooling coils. With the mother liquor solubility data, it was now possible to calculate the D_{SAT} of the mother liquor in contact with the cooling coils. This was calculated for the worst contact scenario (i.e. for the tubes entering the Cooling Crystalliser that carry the coldest cooling water into the vessel). This profile was also added to the process trends (as shown in figure 10). From these early trends, it became apparent that at the seeding point, and during the later stages of the Cooling Crystalliser cycle, the D_{SAT} of the bulk mother liquor ran up to 1.15. At the same points in the cycles, it was shown that the mother liquor next to the inlet cooling water tubes could rise just above 1.2 D_{SAT}.

The limits for D_{SAT} in technical sugar syrups indicate an uncertain transition phase between the metastable and labile zones (see figure 11), potentially at or above a D_{SAT} of 1.15. A recent publication by Lajos Rozsa, Jakab Rozsa & Seppo Kilpien[3] refers to these critical limits being defined by the syrup purity, as described by Broadfoot & Wright [4], which refers to an upper limit of around 1.12 to 1.13 D_{SAT} over the purity range experienced in our Cooling Crystalliser. Therefore, operating with a D_{SAT} of 1.2 at the cooling coil surface presents a significant risk of false grain formation and should be avoided. In figure 12 it is evident that some false grain developed, leading up to the seeding point and

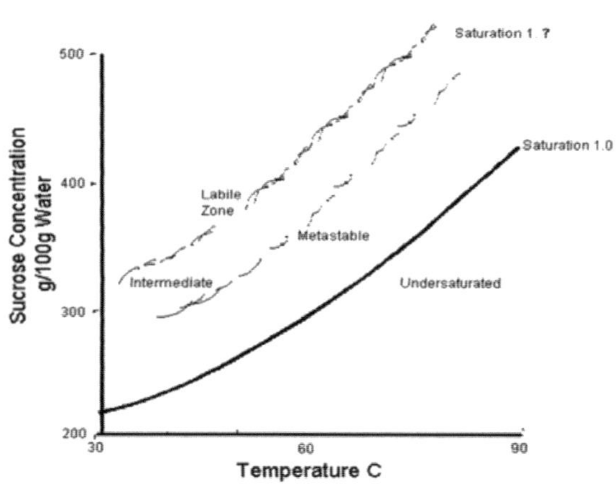

Figure 11 Classic Miers diagram.

also potentially during the later stages of cooling, as seen by the drop in mother liquor brix.

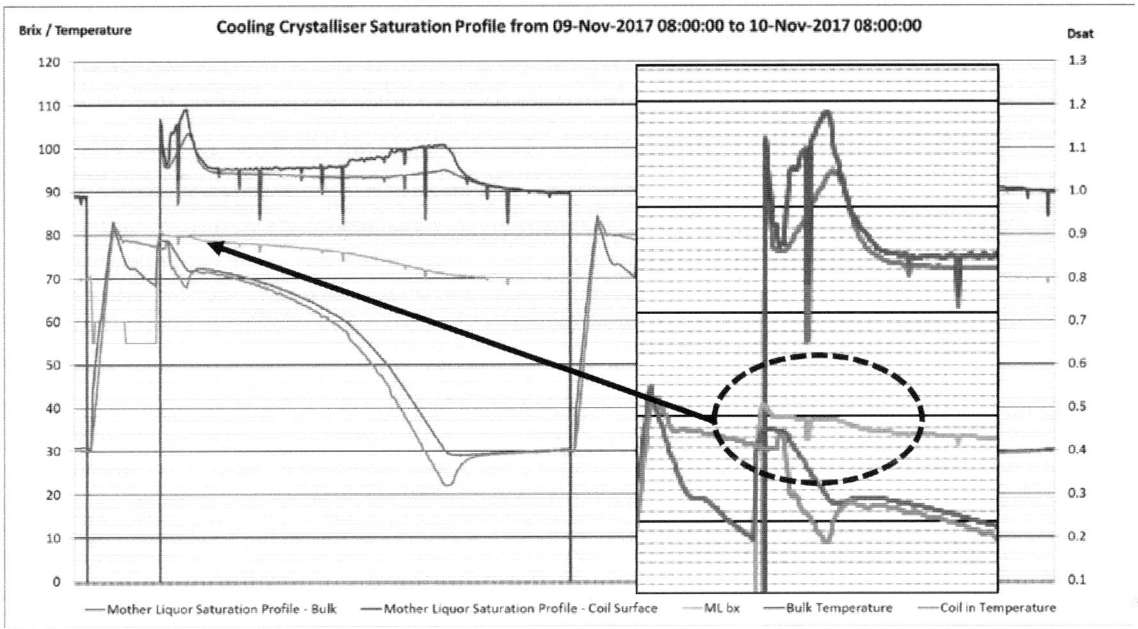

Figure 12 Crystalliser trend showing brix dropping pre-seed addition (highlighted in the elipse)

Based on the new D_{SAT} trends, the cooling profiles were adjusted to ensure that both the cooling coil and bulk liquor temperatures did not force the potential risk of false grain formation during the cooling cycle. An example of the revised trend can be seen in figure 13.

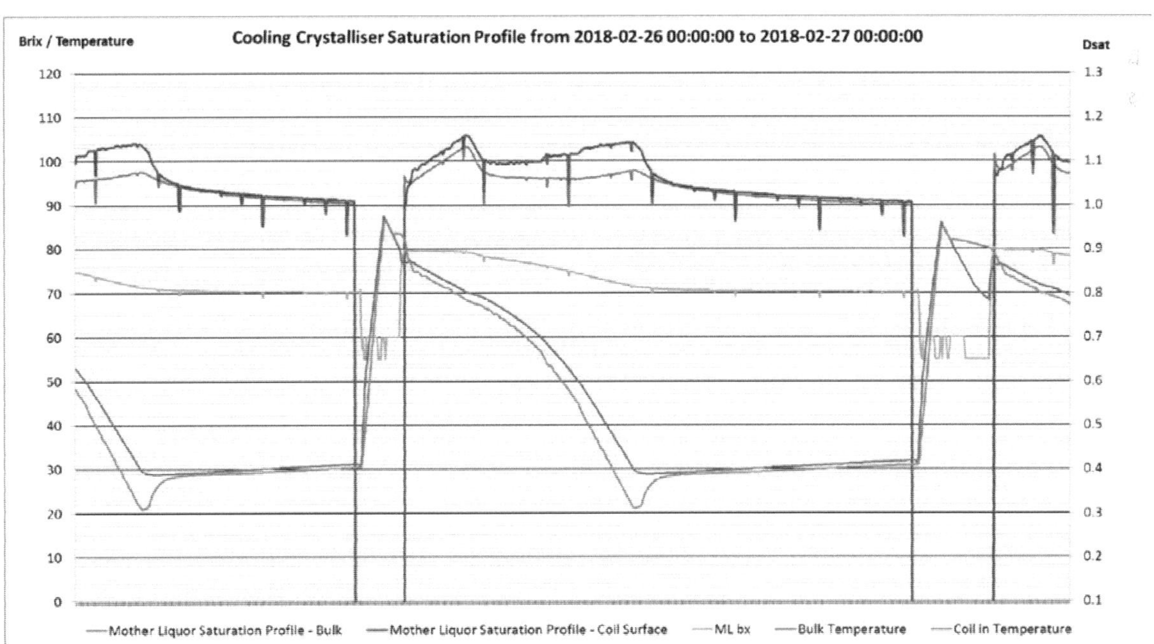

Figure 13 Revised cooling profile - lower D_{SAT} and no brix drop-off pre-seeding (green line)

Instrumentation improvement benefits

- ***Mother liquor monitoring***

The installation of the refractometer into the Cooling Crystalliser unlocked significant potential in controlling the performance of the whole Cooling Crystalliser cycle. Its repeatable and reliable brix output enabled the Wissington team to modify the cooling profile. With an accuracy within 0.2 brix of the laboratory analysis, this allowed the Cooling Crystalliser seed brix target to be consistently met. Its output was used to re-align the cooling profile to improve both the crystal quality (regularity) and yield (crystal content).

Ultimately, by developing the D_{SAT} calculation from the refractometer output, it became possible to manage both the bulk and cooling tube temperatures to minimise the risk of false grain production. This allowed the team to capitalise on the increased crystal content and improved crystal size control.

- ***A link to particle size growth***

Prior to the 2017 Juice Run, a Neltec ColourQ Pan colorimeter was installed on the Cooling Crystalliser. This was the first time the colorimeter had been installed on a Cooling Crystalliser. Whilst the unit is still in development, data from the system has been used to assist with the interpretation of the impact of D_{SAT} on the crystal size development during the cooling profile.

Combining the outputs from both the refractometer and colorimeter instruments has allowed the Wissington team to further modify the cooling profile, whilst monitoring both D_{SAT} and the crystal size growth. This has increased the output capacity of the Cooling Crystalliser, allowing the work on the Linear Pan Growth project to progress without impacting on factory crystal throughput.

Figure 14 highlights one of the observed relationships between the measured particle size and the calculated D_{SAT}. Post seeding, the D_{SAT} falls away rapidly and the particle size increases at the same rate. Then, as the D_{SAT} stabilises at around 1.07, the rate at which the particle size changes slows. This is not an unknown phenomenon, as it is a fundamental part of the linear growth theory. However, the combination of D_{SAT} and particle size change has allowed the team to further optimise the cooling profile.

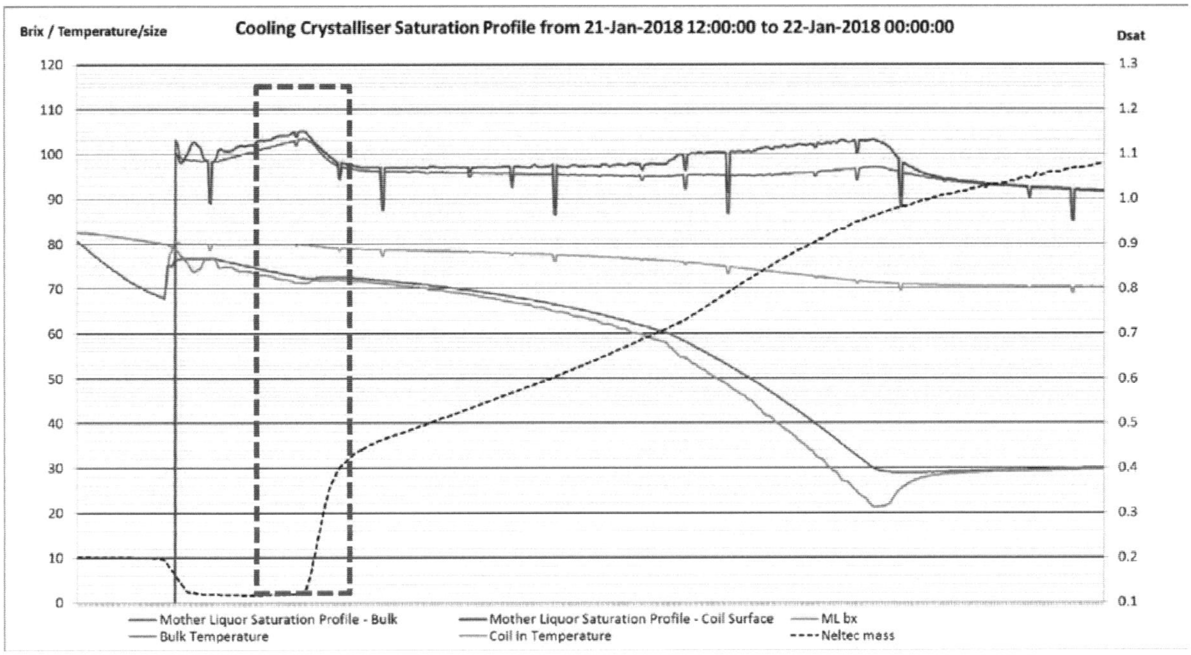

Figure 14 Current Cooling Crystalliser profile showing rapid particle size increase as D_{SAT} falls

Results

There have been a number of significant improvements delivered along the journey in developing the Cooling Crystalliser at Wissington, each one adding to the understanding of the system. These were;

- Ensuring the seed slurry is always added into the vortex created by the stirrer (adjacent to the stirrer shaft), improving its incorporation into the bulk liquor.
- Controlling the slurry addition rate to ensure it is rapidly drawn into the bulk liquor. Adding the additional set of stirrer blades below the surface layer to significantly increase the drawdown of the seed slurry into the bulk liquor

All three changes above resulted in a step reduction in seed massecuite particle size, whilst also contributing to the crystal content increase with the seed massecuite. Following these actions, the project's focus switched to improving the cooling profile. This resulted in the following developments:

- Optimising the cooling profile, based on the data gained from installing the K-patents refractometer, improving particle size (MA) and crystal content.
- Calculating the D_{SAT} from the refractometer data enabling a more targeted cooling profile, based on driving the mother liquor D_{SAT}
- Ensuring that the D_{SAT} on the cooling coils does not exceed the 1.15 D_{SAT} limit.

- Comparing D_{SAT} profiles with the data from the Neltec ColourQ pan colorimeter, increased the understanding of how and when the seed particles were growing.

The final three actions above enabled the team to revisit the cooling profile, ultimately leading to further changes and optimisation of the system. These additional changes delivered an improved crystal content and reduction in overall cycle time, improving the overall capacity of system. Additionally, the repeatability of the crystal particle size was improved; an important factor in controlling the quality of crystal from the batch pans. Figure 15 below shows the culmination of these changes in the current profile.

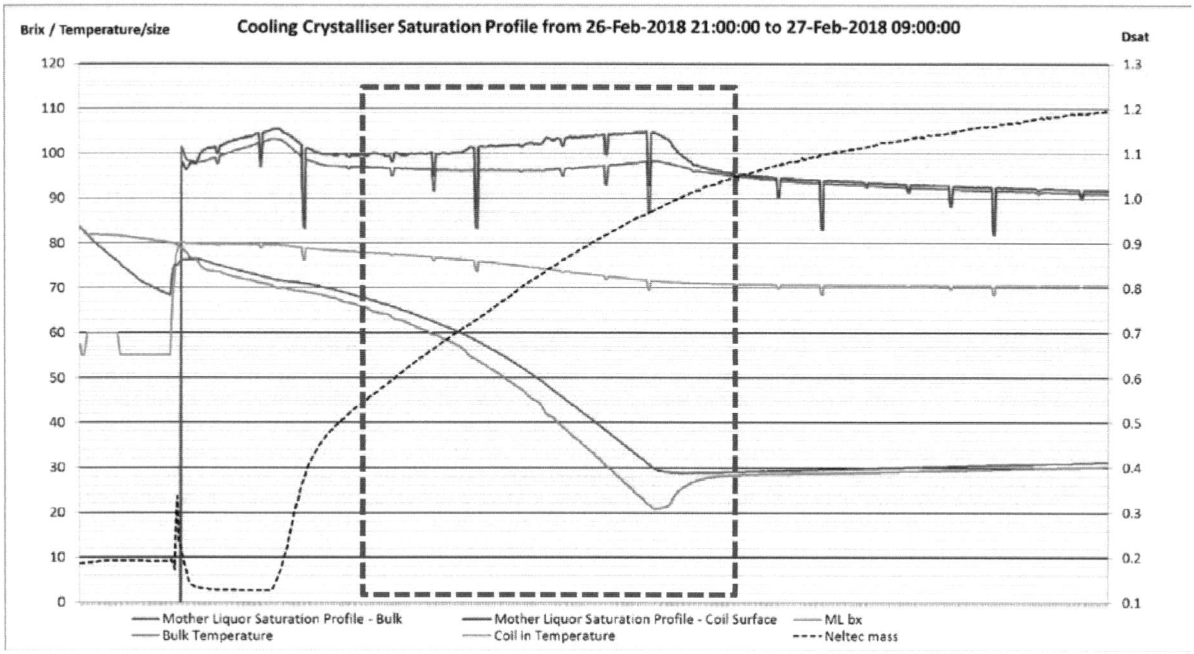

Figure 15 Current cooling profile

Conclusions

Delivering the optimum crystal product, whilst balancing the key factors of product quality control, is reliant on a number of key factors. These all need to be in alignment to produce an even, consistent seed massecuite for the white pans. These factors are:

- Accurate Mother liquor brix measurement – a key factor in controlling and optimising the D_{SAT}, and hence maintaining optimal growth rate conditions to deliver a consistent, quality product.
- A cooling profile that holds the D_{SAT} at the best conditions for crystal growth, whilst ensuring the conditions at the surface of the cooling coils do not rise above 1.15 DSAT, to prevent false grain occurring.

- Delivering the seed slurry to the crystalliser at the right location, within the well mixed zone, close to the stirrer shaft, at the right addition rate to reduce seed clumps on the liquor surface.
- Designing the stirrer configuration to quickly draw the seed slurry into the bulk liquor, before the carrier fluid can evaporate, ensuring the seed crystals are dispersed evenly throughout the bulk liquor.

The outcome from the trial work at Wissington has produced a best practise for replication across the other British Sugar sites, based on;

- Installing a refractometer in the Cooling Crystalliser,
- Modifying the cooling profile to manage the D_{SAT} profile of both the bulk liquor and the coil surface liquor,
- Reviewing the stirrer design for the upper layer control,
- Relocating the seed inlet to align with the surface layer vortexing,
- Regulating the slurry flow into the vessel to match the rate of its incorporation into the bulk liquid.

Based on these factors, seed size regularity and crystal content from the Cooling Crystalliser system can now be relied upon to deliver a consistent MA, which is tuneable by the operators through transfer brix settings, or through changes to the slurry addition volume.

Acknowledgements

The authors would like to thank the Wissington factory operational team for their support and resources used during the trial

References

1. Development of a Linear Pan Growth Approach, Chris Mayhew, Carlton Haynes, Robert Howe, Craig Parker and Chris Rhoten, British Sugar and Escon America Inc. SIT Proceedings from the 76th Technical Conference (Taiwan) 2017

2. Linear Growth Rate Based Adaptive Batch Crystallisation Control, Christopher D Rhoten, ASSBT Proceedings from the 38th Biennial Meeting, February 23 - February 27, 2015

3. Rozsa, L., Rozsa, J., Kilpinen, S., 2017, 'Crystal growth and crystallization control tactics in industrial sugar crystallizers, Part 2: crystal growth', International Sugar Journal, pp260-266

4. Broadfoot R., Wright P.G., 1972, Nucleation Studies, Proc. 39th Conf. QSSCT; 353-362

S.I.T. Paper #1182

Powder characteristics of granulated sugar in Japan

Miki Sakazaki, Kohei Miyazaki, Sayoko Fujii,

Yukie Nagai, Toshikazu Kawai and Akio Hirata

Research Section, Research and Development Division, Mitsui Sugar Co., Ltd.

36-2, Nihonbashi-Hakozakicho, Chuo-ku, Tokyo, 103-8423, Japan

TEL+81-3-6758-6166

E-mail: Miki.Sakazaki@mitsui-sugar.co.jp

Introduction

The caking of sugar is a major problem in Japan. Our company has received on average of 157 complaints about caking (Figure 1). This would be considered to be due to Japan having four distinct seasons with extreme climate changes, including changes in humidity and temperature levels (Figure 2). It also seems that customers have become sensitized to the quality of products. The caking of sugar makes powder harder to handle and lowers manufacturing efficiency. Consequently, there is need to investigate the factors contributing to caking to prevent it.

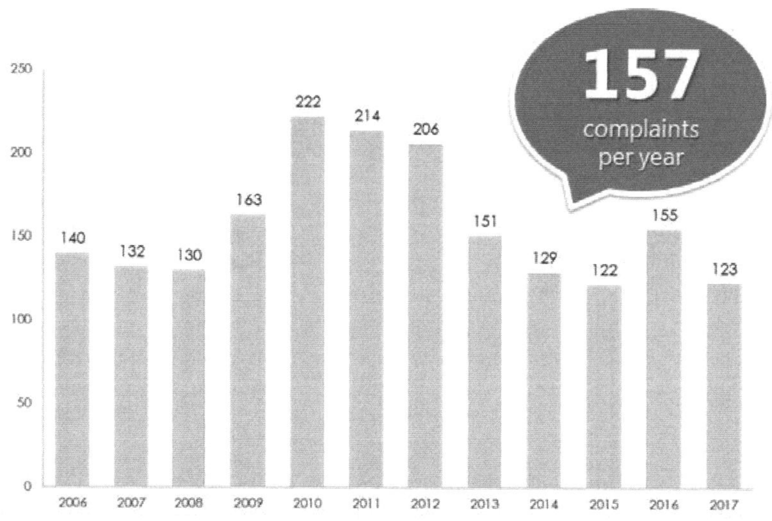

Figure 1. Complaints about caking (2006-2017)

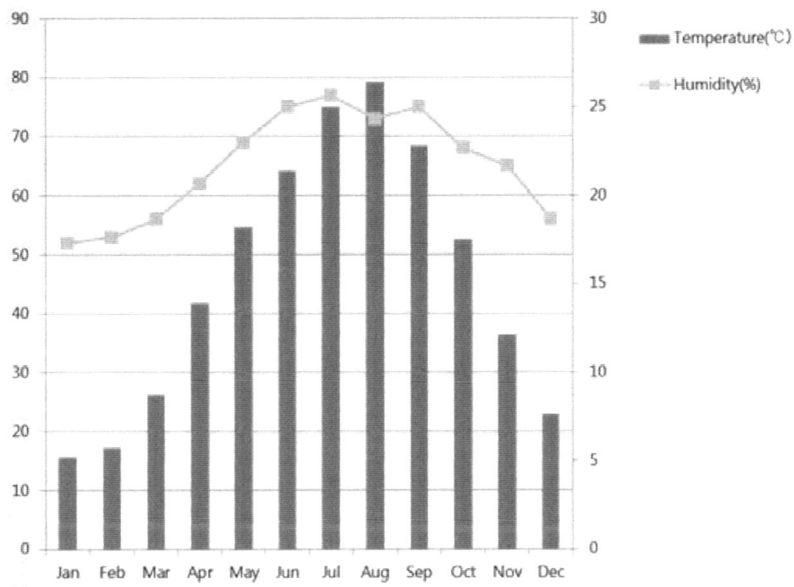

Figure 2. Temperatures and humidity for Japan (1987-2010)

In general, the caking of sugar is caused by environmental changes, namely changes in humidity levels, the dissolution and recrystallization of crystal surfaces, and the adhesion of crystals. The crystal goes through four different stages during the caking phenomenon [(2)] (Figure 2).

These four stages are:

A : The pendular stage at which the sugar is still free flowing,

B : The funicular stage corresponding to the establishing of permanent contact between grains,

C : The capillary stage reached when the moisture level is high provoke liquid bridges between grains,

D : The drop stage obtained when the dissolution of grains is preponderant.

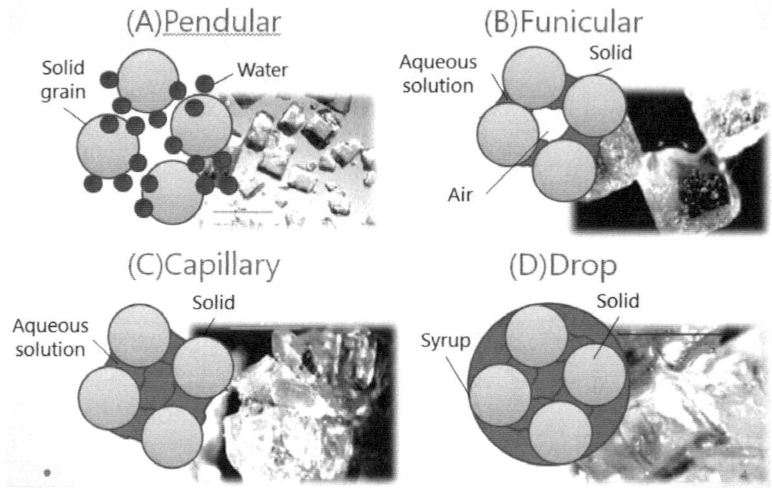

Figure 3. 4 stages of caking (Gutman, 1992)

Many complicated factors are related to the caking phenomenon, but they can be broadly classified into two groups: internal factors and external factors. The internal factors are powder characteristics, such as grain size, flowability and crystal surface property, and water activity (or moisture content). External factors are packaging condition, storage environment (humidity or temperature), storage time, and pressure during storage among other things. Some of these factors are why caking is a big problem, especially in Japan.

Evaluating the caking of sugar quantitatively is necessary for taking measures against. However, it is very difficult to establish the method to evaluate caking phenomenon physicochemically. Based on a previous study[4], establishing the method has 4 main components as follows: (1) to measure powder characteristics related to caking, (2) to adjust caking of sugar, (3) to measure the physical properties corresponding to the strength of caking, (4) to evaluate caking quantitatively. However, little has been reported on the powder characteristics of Japanese sugar, compared to those of salt or other food powders [3],[7]. This study investigated the powder characteristics of granulated sugar in Japan, the first of the four mentioned components.

Materials and Methods

Materials

This study used granulated sugar with its simple surface texture. Samples of granulated sugar were collected from 19 products which were different in types of grain sizes (Table 1):

- 4 granulated sugar products from (GS: 500-650 μ m): 3 of our products (Mitsui①-③), which were from factories located in different regions of Japan, and a company A product were used.

-15 fine granulated sugar products (FGS: 200-500 μm): 4 of our products (Mitsui①-④), one product from each of companies A-E, 4 products from company F (F①-④) and 2 products from company G(G①-②) were used.

Table 1. Studied products

	No.	Product	
GS (500-650μm)	1	Mitsui	①
	2		②
	3		③
	4	Company A	
FGS (200-500μm)	5	Mitsui	①
	6		②
	7		③
	8		④
	9	Company A	
	10	Company B	
	11	Company C	
	12	Company D	
	13	Company E	
	14	Company F	①
	15		②
	16		③
	17		④
	18	Company G	①
	19		②

Methods

Crystal properties

The crystal shape and surface properties were observed under Miniscope® TM3030 (Hitachi High-Technologies corp.), a low-vacuum electron microscope, and Digital Microscope KH-7700 (HIROX Co., Ltd.).

Grain size

The average grain size, grain size distribution and uniformity were measured by Mastersizer 300 (Malvern Instruments, Ltd.), a laser diffraction scattering type grain size measuring device.

The average grain size is expressed in the median diameter, the value of the grain diameter at 50% in the cumulative distribution (D_{50}). The uniformity, which is one of the indicators of grain size distribution, is calculated by the diameter at 60% (D_{60}) divided by the diameter at 10% (D_{10}). (In generally, when the D60 and D10 would be close in value, the powder would not be flocculated.)

Powder characteristics

The repose angle, compression degree, spatula angle and flowability were measured by Multi tester MT-02 (Seishin Enterprise CO., Ltd.). The repose angle is the slope at which the powder will remain in place without sliding. The compression degree expresses the freely settled density (ρ_A) and the tapped bulk density (ρ_T) of the powder. It is calculated by this formula: $(\rho_T-\rho_A)\times 100/\rho_T$. The spatula angle is the slope angle of the powder, which measured by inserting a spatula into the powder parallel to the bottom of the container and pulling it out. Carr's flowability index (Carr index) is used as a comprehensive index for the flowability of a powder. It is calculated from four measured values (repose angle, compression degree, spatula angle, uniformity) and converted to a scale of 25 for each, making for a total of a hundred. (A Carr index greater than 90 is considered to be an index of good flowability, while one of below 20 is considered to be poor flowability.)

Statistical processing

The correlation coefficients of the measured values were determined by BellCurve for Excel (Social Survey Research Information Co., Ltd.). Those measured values were analyzed using principal component analysis (PCA) to group products.

Results

Microphotography pictures

Microphotography pictures allow the crystal surface properties, which affect powder characteristics, to be understood. There were not any clear differences between the four GS products. However, minor differences were recognized between our products (Mitsui①-③) and A. Picture 1 in figure 4 and 5 shows Mitsui's standard product. In picture 2, the crystal surface of A was smoother and the edges were sharper than Mitsui's product.

Figure 4. Digital microphotography pictures of GS (Mitsui③ and A)

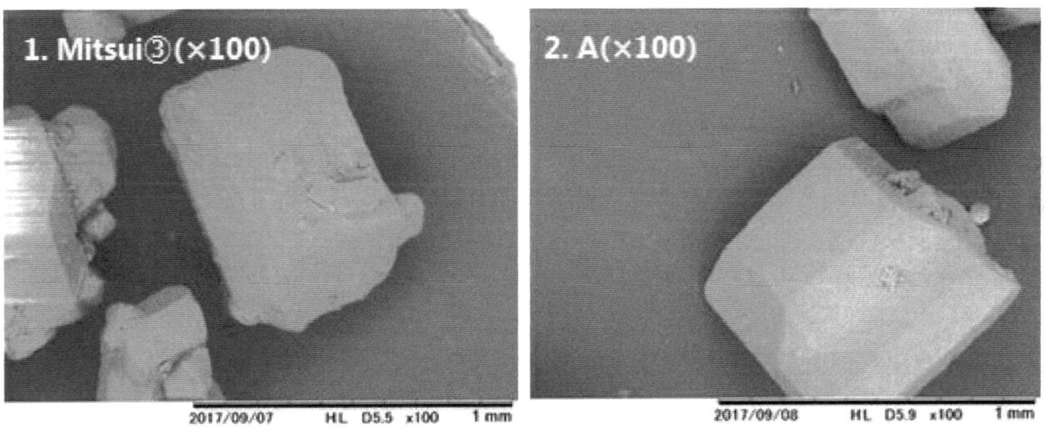

Figure 5. Microphotography pictures of GS (Mitsui③ and A)

Regarding FGS products, irregular crystal shapes were observed in Mitsui① and F③. In addition, more fine particles were seen than other FGS products such as G① (Figure 6).

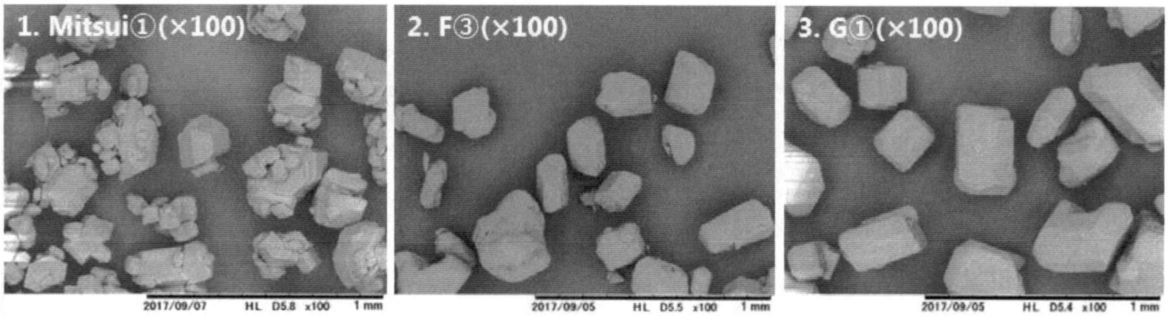

Figure 6. Microphotography pictures of FGS (Mitsui①, F③ and G①)

Powder characteristics

Each of the measured values is presented in Table 2. Comparing GS and FGS products, compression degrees tended to be smaller, and Carr indexes of GS tended to be larger than with FGS. Regarding the FGS products, in particular, it was found that F① and F② products had low spatula angles and large Carr indexes. On the other hand, grain sizes were small and compression degrees were large on the Mitsui① and F③ products with small Carr indexes, namely with poor flowability.

Table 2. Measured values of each product

	No.	Product	Grain size(μm)	Repose angle(°)	Compression degree(%)	Spatula angle(°)	Uniformity	Carr index
GS	1	Mitsui ①	616	32.8	10.1	12.3	1.8	91.5
	2	Mitsui ②	555	32.4	12.2	12.8	1.8	90
	3	Mitsui ③	599	31.8	8.4	12.2	2.1	92
	4	A	553	32.7	8.5	31.5	2	89
FGS	5	Mitsui ①	224	33.8	18.9	49	2	78
	6	Mitsui ②	373	30.9	10.5	39.8	1.9	85
	7	Mitsui ③	298	28.2	13.5	43.3	1.8	86
	8	Mitsui ④	305	33.4	15.6	44	2.2	81.5
	9	A	282	34.1	15.4	54	1.9	80
	10	B	352	26.8	14.1	40.8	1.8	86
	11	C	305	31.6	13.1	41.8	1.8	83
	12	D	416	36.8	11.4	50.7	1.8	79
	13	E	348	38.3	12.1	52.9	1.8	78
	14	F①	304	28.6	13.5	12	1.7	93
	15	F②	282	31.4	17	16.4	2.3	88
	16	F③	236	31.8	18.4	46.6	2	78
	17	F④	346	35	12.3	51.7	1.6	80
	18	G①	308	28.4	14.3	43.4	1.7	86
	19	G②	254	33	16.5	45.1	1.8	79.5

Statistical processing

The correlation matrix indicating a relationship is shown in Table 3. The following three relationships were found to have a strong correlation ($p<0.01$): grain size and compression degree, repose angle and Carr index, and spatula angle and Carr index.

Table 3. Correlation matrix of each variable using PCA

	Grain size	Repose angle	Compression degree	Spatula angle	Uniformity	Carr index
Grain size	1.00	0.17	-0.90**	0.10	-0.37	0.11
Repose angle	0.17	1.00	-0.09	0.53	0.07	-0.76**
Compression degree	-0.90**	-0.09	1.00	-0.08	0.56	-0.21
Spatula angle	0.10	0.53	-0.08	1.00	-0.25	-0.86**
Uniformity	-0.37	0.07	0.56	-0.25	1.00	-0.05
Carr index	0.11	-0.76**	-0.21	-0.86**	-0.05	1.00

*P<0.01

Every measured value was subjected to PCA to acquire first to sixth components. The eigenvalues for the components were 2.468, 2.281, 0.821, 0.332, 0.074 and 0.025. First and second principal components (PC) having large contribution rates were employed (Table 4).

Table 4. Eigen value and contribution ratio of each principal component using PCA

	Eigenvalue	Contribution rate	Cumulative contribution rate
PC1	2.468	41.13%	41.13%
PC2	2.281	38.01%	79.14%
PC3	0.821	13.68%	92.82%
PC4	0.332	5.53%	98.36%
PC5	0.074	1.23%	99.59%
PC6	0.025	0.41%	100.00%

As a result of the principal component analysis, the repose angle, spatula angle and Carr index had large principal component load quantities for PC1. On the other hand, the grain size and compression degree had large load quantities for PC2 in the following table5 and figure7.

Table 5. Principal component load quantities of each measured value using PCA

	Grain size	Repose angle	Compression degree	Spatula angle	Uniformity	Carr index
PC1	0.3148	0.8199	-0.2642	0.8902	-0.2577	-0.8763
PC2	-0.8462	0.1905	0.9328	0.1374	0.6541	-0.4597

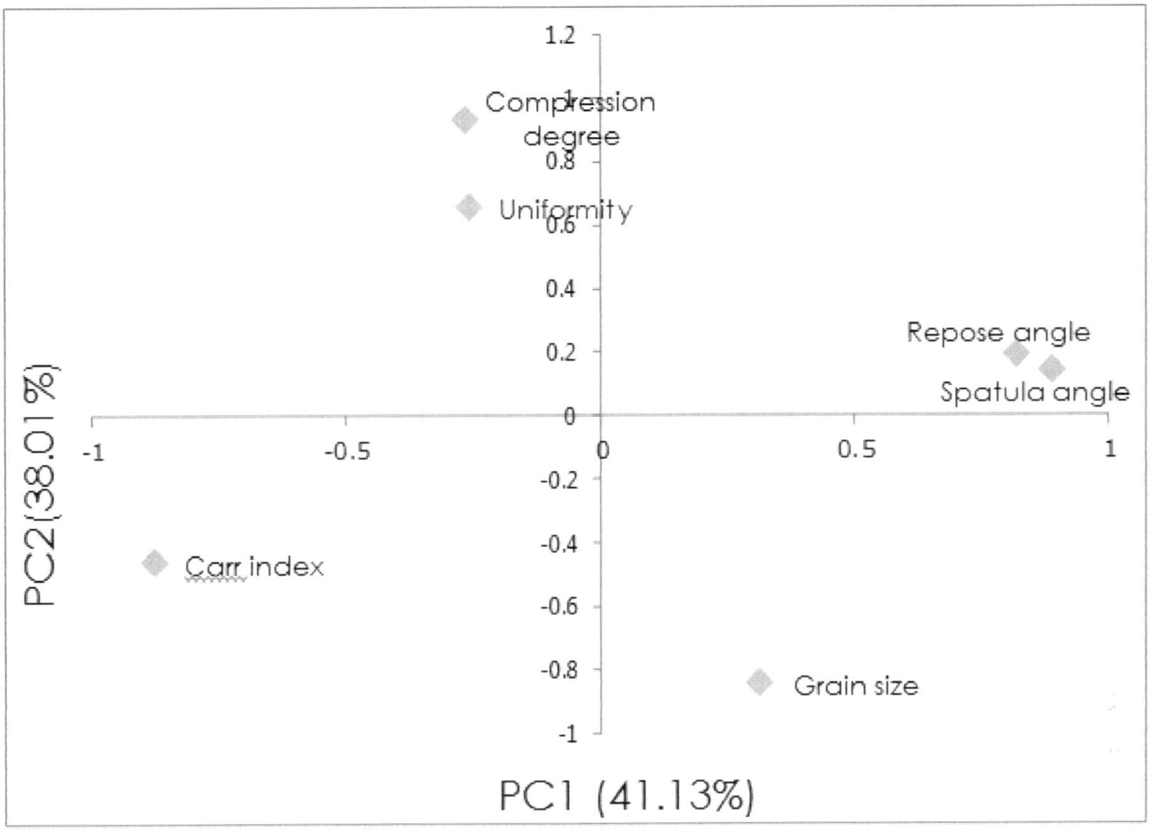

Figure 7. Loading plot using PCA (PC1 vs. PC2)

The principal component scores of each of the products (PC1 vs. PC2) are presented in Figure 8. F① and F② products had large scores in negative directions of PC1. The other products were divided into the following three groups: positive scores of PC1 and positive scores of PC2, negative scores of PC1 and negative scores of PC2, and positive scores of PC1 and negative scores of PC2.

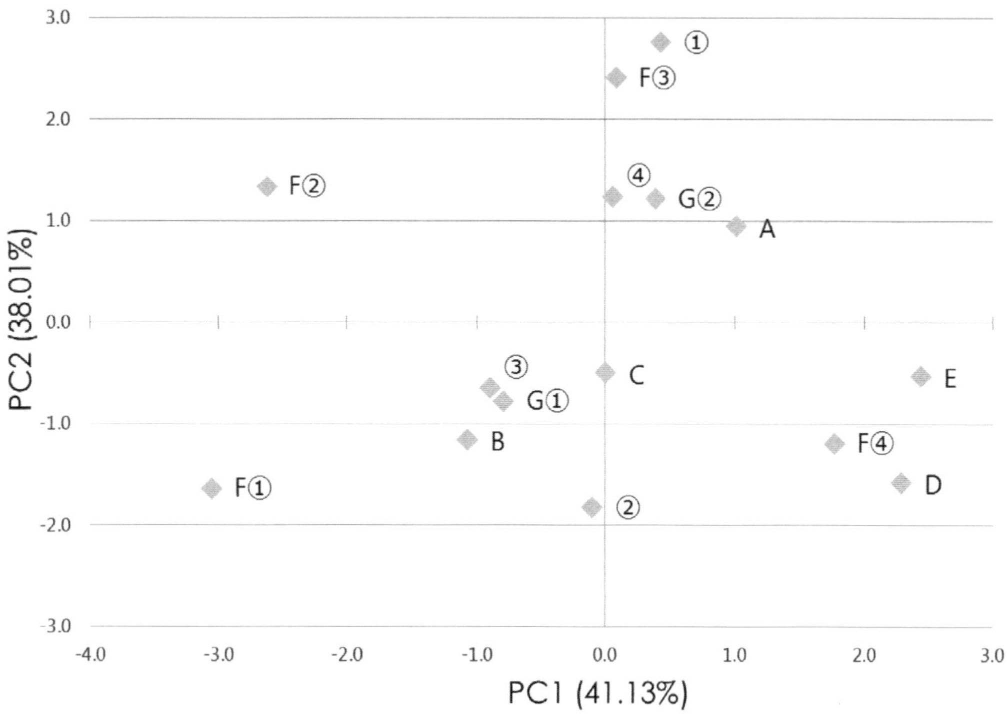

Figure 8. Principal component scores of each product using PCA (PC1 vs. PC2)

Discussion

The present study is the first step to evaluate caking quantitatively. To the best of our knowledge, there is no data which compares different granulated sugar products in Japan.

Carr index is used as an index of flowability of a powder, and flowability of sugar is reduced when it cakes. Therefore, it was expected that the products with small Carr indexes tend to easily cake. However, as a result of our study, there weren't significant differences as a whole. One difference is that FGS products were tended to be smaller than GS. It is assumed that when the grain size is small, flowability would be poor because contact area of crystal surfaces is large and have more inter-grain resistances.

The principal component analysis of FGS was aimed to clarify the tendency of the products. It was found that PC1 was explained by the repose angle, spatula angle and Carr index, and PC2 was the grain size and compression degree. At the same time, the repose angle and spatula angle have a correlation with Carr index. Hence, it seems that PC1 is represented by flowability. However, a comprehensive description of PC1 could not be found because it comprised positive and negative directions for the principal component load quantities. The reason for this is that the large repose and spatula angles express the poor flowability, while on the other hand, the large Carr index expresses good flowability. In the same way, the grain size would be the most important indicator of PC2, since the grain size and compression had a strong correlation. Consequently, these results suggested that flowability and grain size determine the powder characteristics of granulated sugar.

As a result of the principal component scores of each of the product, F① and F② products had different tendencies from other products, and they had large scores in the negative directions of PC1. This would be affected by extremely low spatula angles of the products. Other products could be divided into three groups because of their distinctive principal component scores.

As a result of comparing crystal properties using microphotography pictures, there were some differences in crystal shapes, especially of FGS products. The crystal shapes of Mitsui① and F ③ products, with their small grain sizes and large compression degree, were characterized as non-crystal particles. These fine particles play the role of binders with crystals, and they have an adverse influence on flowability. Therefore, crystal property would be one of the factors in powder characteristics.

However, more study is needed to confirm these findings. As a next step, we will further consider the correlation between the measurement of caking intensity and these powder characteristics.

Conclusions

In conclusion, the present study confirmed the following:

- It was found that powder characteristics of granulated sugar different among products.
- Granulated sugar with large grain size has great flowability.
- Irregular fine particles were observed in products with a small grain size and low flowability.
- Flowability and grain size are important indicators of the powder characteristics of granulated sugar.
- There is a possibility that crystal properties affect powder characteristics.
- The correlation between these powder characteristics and caking will be researched.

References

1. Barbara R. and Mohamed M. 2003. Caking of white crystalline sugar. International sugar journal. 105(1251). 128-136

2. Gutman S. 1992. Contribution à la connaissance physico-chimique des poudres: la notion d'humidité relative d'équilibre. Thèse de Doctorat en Pharmacie, UFR de Pharmacie Chatenay - Malabry: Paris XI.

3. Ito H. and Shinohara T., et al. 1999. Method for preventing solidification of salt. JP3638399A. 1999-02-15.

4. Kanazawa T. 1970. Caking phenomenon of powder and measures against caking. Gypsum & Lime. 106. 29-34.

5. M.Mathlouthi and B Roge. 2004. Caking of white sugar and how to prevent it. *Proc S Afr Sug Technol Ass*. 78. 495-504.

6. Pakamon C. and Jirawate C., et al. 2006. Effects of grain size, reducing sugar content, temperature and pressure on caking of raw sugar. *Kasetsart J. (Nat. Sci.)* 40. 141-147

7. Takeda K, and Takeda T., et al. 2013. Method for preventing powdered food product from solidifying, and powdered food product. JP2013269615A. 2013-12-26.

USE OF MULTIFUNCTIONAL ADSORBENTS AS A PRETREATMENT IMPROVES GAC COLUMN PERFORMANCE

Idalberto Delgado. Sugar Applications Engineering Manager. Sugar Segment.
Jerry Lengen. Business Development Manager, Sugar Segment.
Mario Canales. Sugar Applications Engineer. Sugar Segment.
Bruce Belcher. VP/GM Sugar segment.

Graver Technologies; 200 Lake Drive; Glasgow, DE 19702 [USA]

Summary

Insufficient column life between regeneration cycles is often a problem in refineries utilizing regenerable columns, whether Granulated Active Carbon (GAC), Ion Exchange Resin (IER), or Bone Char (BC). As a result, the color after decolorization is often higher than desired, leading to reduced operating rates through the columns and/or reduced refined sugar yields. Lower color and/or turbidity in the liquor feeding the regenerable columns extends their operating cycles and improve their capacity utilization.

Pretreatment with Ecosorb® multi-functional (non-regenerable) adsorbents can significantly lower both color and turbidity, resulting in improved column efficiency, increased yields, and/or lower refined sugar color.

A full-scale plant trial was conducted using Ecosorb® S-492 as a pre-coat on the secondary (polish) filters after carbonatation and prior to GAC treatment, at Taiwan Sugar Corporation. The S-492 pre-coat reduced the color of the clarified liquor by 40% and increased the overall color removal efficiency from the First Filtrated Carbonated Liquor to the Decolorized Liquor feeding evaporation/crystallization processes from approximately 70% to 90%, with no adverse effect on flow rates or pressure drop across the secondary filter.

As a result, Taiwan Sugar Corporation realized two major benefits:

1. the refinery increased his refined sugar yield by decreasing time washing procedures to reduce losses while producing the same refined sugar quality as before, or
2. by keeping the washing procedures as before, refinery produced a lower color refined sugar, which may be of value in certain markets.

The economic and operational advantages of reducing washing in centrifuges can be quite large, resulting in both increased refined sugar production, and energy savings, since less run-off is generated, reducing both steam and power consumption required to re-crystallize the run-off.

Background

Ecosorb® multi-functional adsorbent products have been used for more than twenty years to produce high quality, low color liquid sugar from refined crystalline sugar. Ecosorb is known to be a versatile and robust technology and can be used in a variety of ways to improve the over-all efficiency of refining operations.

The general mode of operation was originally as a pre-coat, which involves loading all the adsorbent on the filter at the beginning of the filter cycle (after the initial filter aid pre-coat), achieving all the color reduction as the sugar solution passes through the filter cake.

This pre-coat operation is possible due to the combination of the rapid kinetics of small adsorbent particles combined in an open, highly permeable structure (Delgado and Lengen, 2016). This unique combination allows the use of thicker filter cakes with low pressure drop for maximum decolorization and extended filter cycle life (up to 1 week, depending on the service).

Over the years Ecosorb® applications evolved to include treatment of higher color sugars, in batch slurry, continuous slurry, and hybrid slurry/precoat modes of operation. All these modes of operation take the advantage of the significant "press effect" exhibited with proper use of Ecosorb® adsorbents.

"Press effect" (realizing full adsorptive capacity of the media) is defined as the residual adsorptive capacity of the adsorbent in the filter cake. The magnitude of the "press effect" depends on the type and amount of adsorbent accumulated on the filter; the thickness of the filter cake; and the flow rate (Delgado and Lengen, 2016), which together determine the contact time and ultimately the mass transfer capability in the filter cake).

While a very modest "press effect" is sometimes seen in a standard Powder Activated Carbon /Diatomaceous Earth (PAC/DE) adsorptive process, the "press effect" with Ecosorb® adsorbents is much more pronounced (Delgado and Lengen, 2016). As a result, in batch operations, the residual capacity of the adsorbent from the first batches of a filter cycle contribute to the decolorization of later batches, allowing the use of much lower dosages on the later batches. "Press effect" is significant starting at Ecosorb dosages of 1-2 kg/m^2 (0.2-0.4 lb/ft^2) and loadings of 5-10 kg/m^2 are typical.

Over the past several years, a new generation of Ecosorb® products has been introduced as a replacement for PAC as the primary decolorizing technology in refineries utilizing non-regenerable adsorbents rather than IER, GAC, or BC. In addition, most refineries using non-regenerable adsorbents use multiple filters running in parallel. Due to the fact filter cleanings are usually staggered, it is possible to have different press effect when filters operate staggered too. Therefore, when Filter 1 is new, and exhibiting very little press effect, Filter 2 is partially full, starting to experience some press effect, and Filter 3 is nearing the end of its filter cycle, showing maximum press effect. This averaging effect allows the refiner to keep the Ecosorb® dosage constant as long as the feed color and fine liquor color target remain fairly constant.

Because of the unique properties of Ecosorb® adsorbents, particularly the higher flow rates with lower pressure drop, resulting in much longer filter cycle life, it became apparent there may be new opportunities to use Ecosorb®. More specifically, in conjunction with other decolorizing processes to improve the over-all refinery operations and/or allow the use of higher color feedstocks, and/or enable the production of lower color refined sugar, and/or increase yields and reduce losses. Thus, we have chosen to focus this paper on the beneficial use Ecosorb in conjunction with a granulated activated carbon column (GAC).

Methods

Taiwan Sugar Refinery is a traditional carbonatation refinery that produces 1,000 ton of refined sugar per day. The primary decolorization technology used is GAC.

Ecosorb® S-492 (S-492) using in a continuous slurry addition, or in a hybrid partial pre-coat plus a small dosage as a continuous slurry addition as operation mode were firstly considered as the best option. However, the available existing buffer tank to add the absorbent for a continuous slurry addition was quite small, without any agitation system and providing less than 10 minutes contact time. Therefore, it was decided to use S-492 as a pre-coat.

Block diagram in figure 1 describes the refined sugar production process and S-492 addition point during the trial.

Fig. 1. Process Flow diagram, Taiwan Sugar Corporation Refinery.

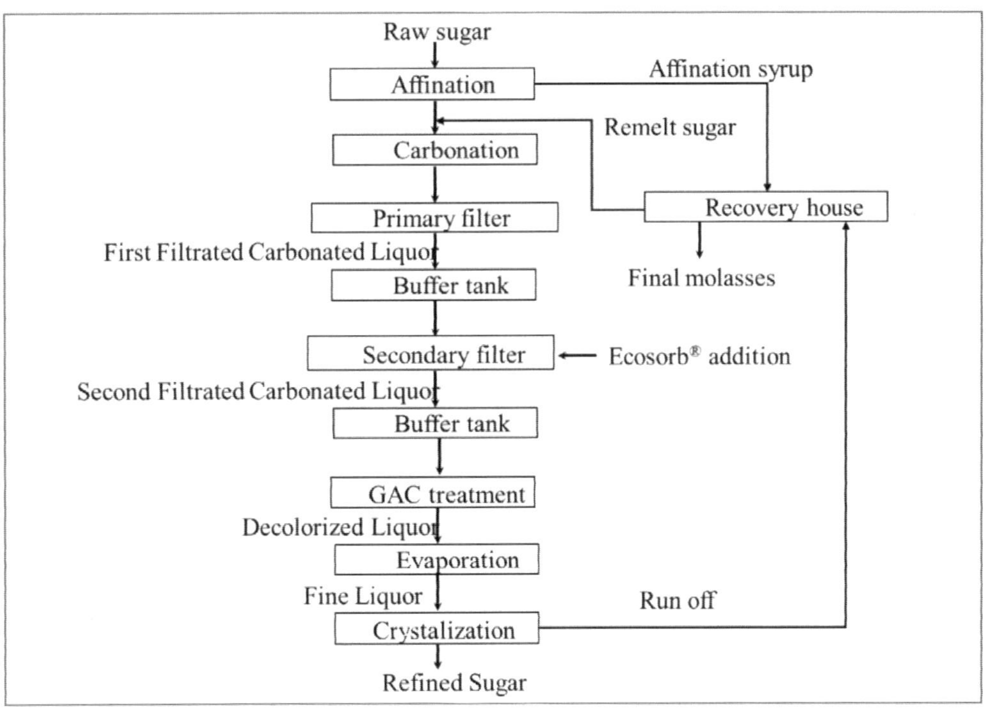

Carbonated Liquor from the primary filter at an average flow rate of 70-80 m^3/hr, fed three 175 m^2 pressure leaf polish filters (average flow rate between 25-30 m^3/hr throughout each filter (2 -3 lpm/m^2). The refinery ran on 2 filters in parallel while a third filter was cleaning.

Filter #1 preparation, Day 1:

- a pre-coat of 100 kg (0.57 kg/m^2) of filter aid was prepared.
- On top of this, a second pre-coat by mixing 750 kg of S-492 (4.30 kg/m^2) and 200 kg (1.14 kg/m^2) of filter aid was prepared.

Filter #2 preparation, Day 2:

- a pre-coat of 100 kg (0.57 kg/m^2) of filter aid was prepared.
- On top of this, a second pre-coat by mixing 750 kg of S-492 (4.30 kg/m^2) and 200 kg (1.14 kg/m^2) of filter aid was prepared.

Filter #3 preparation, Day 3:

- a pre-coat of 100 kg (0.57 kg/m^2) of filter aid was prepared.
- On top of this, a second pre-coat by mixing 750 kg of S-492 (4.30 kg/m^2) and 200 kg (1.14 kg/m^2) of filter aid was prepared.

Each filter was prepared the same on separate days to give an over-all dosage of 0.750 kg absorbent per ton of sugar when steady state was reached. Filter´s pre-coat preparations were staggered to operate the filters the same way to obtain a Second Filtrated Carbonated Liquor mixture at a desirable average color by mixing the filter outlets all the time. This way, a new filter was prepared every 24 hours and filtration cycles for each filter lasted 72 hours.

Results and discussion

Figure 2 shows color removal during the Ecosorb pre-coat operation. Three filters were operated in parallel and staged in such a way that when one filter was new (for example filter 3 at 55 hours in graph) and exhibiting the highest color removal, the absorbent in another filter was partially exhausted, showing a middle color removal (filter 2), and the third filter was nearing the end of its filter cycle (filter 1), with a minimum color removal.

Fig. 2. Color removal over Time

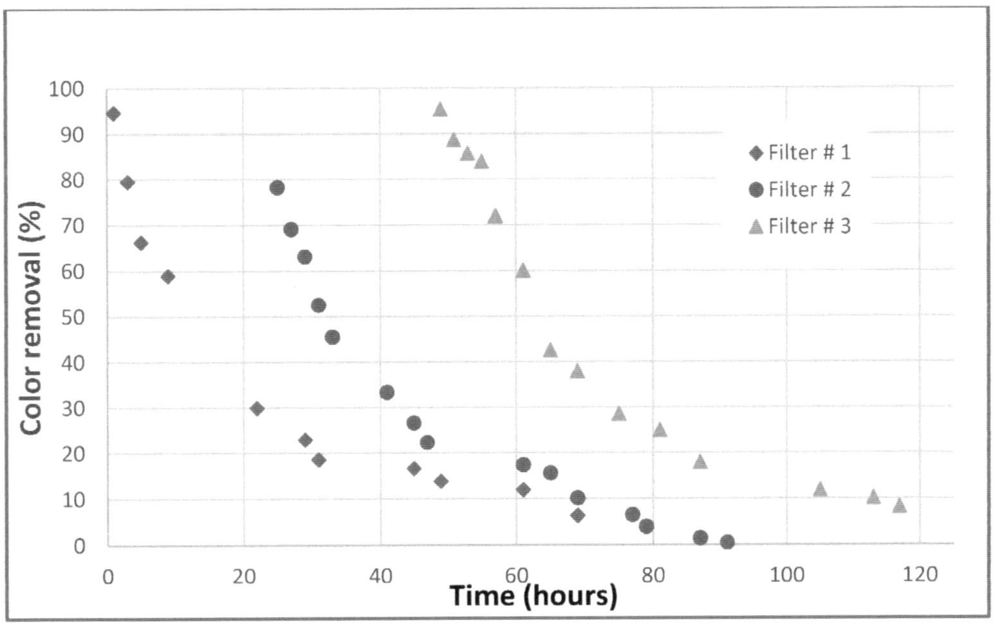

First Filtrated Carbonated Liquor after primary filtration having colors ranging between 600 and 1336 IU were efficiently decolorized with an average color removal of about 40 % through the precoat with a total Ecosorb S-492 loading of 4.3 kg/m^2 in the precoat (a thick enough filter cake for efficient "press effect", about 14 mm) to maximize precoat decolorizing efficiency. (figure 3).

Fig. 3. Stepwise color removal during Carbonated Liquor Treatment from Primary Filtration to Decolorized Liquor.

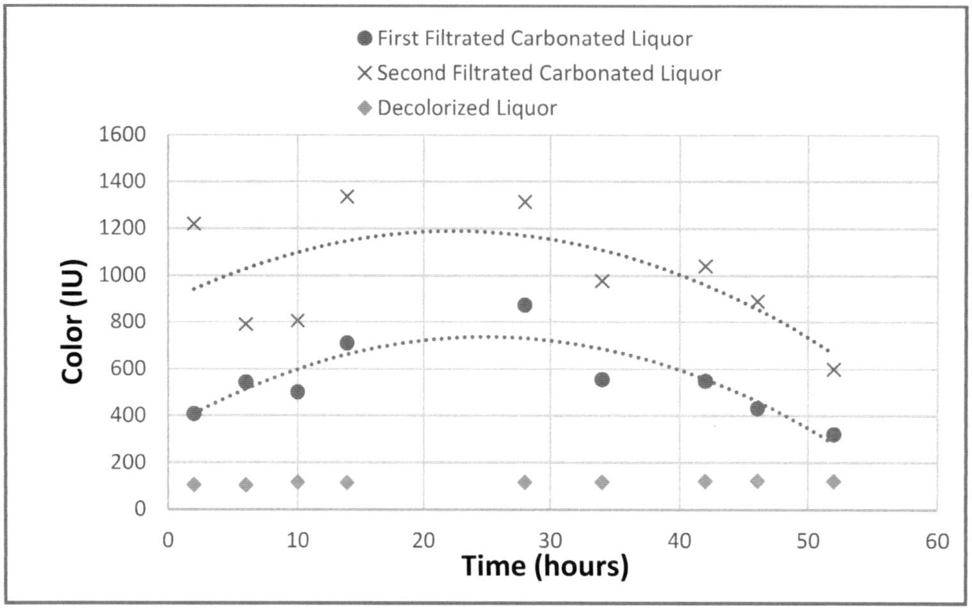

Colors after the Ecosorb precoats ranged from 322 to 874 IU. By combining the filtrate from all 3 precoat filters the average color feeding the GAC was relatively stable, (figure 3). The subsequent GAC treatment yielded a Decolorized Liquor with an average color of 115 IU feeding the crystallization process. Total color removal, including both Ecosorb S-492 and GAC treatments, averaged around 90% regardless of liquor color after carbonatation (figure 4).

Fig. 4. Carbonated Liquor: Color Reduction Across Ecosorb Precoat and GAC Treatments

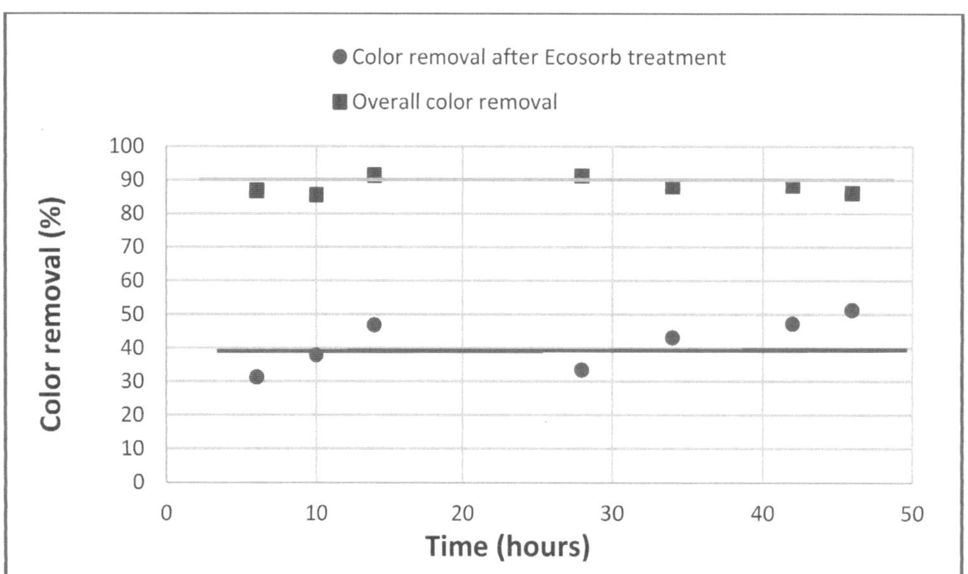

Filter cycles were terminated after 72 hours due to diminished color removal (figure 5) rather than reduced flow or excessive pressure drop. No pressure drops higher than 3.5 bar were observed in any of the filter cycles (figure 5). Despite the high adsorbent and DE loading on the filters (6 kg/m^2 per pre-coat on an as is basis) no excessive pressure drop was observed for two reasons. First, the solids content of S-492 is only about 50% which means that on a solids basis, the Ecosorb dosage is only half of the "as is" adsorbent dosage used, and second, the excellent filtration properties of the Ecosorb allows processing at high flow rate at with low differential pressure across the filter, even with a relatively high solids loading of Ecosorb and DE.

Fig. 5. Flow Rate and Pressure Drop During S-492 Precoat Runs

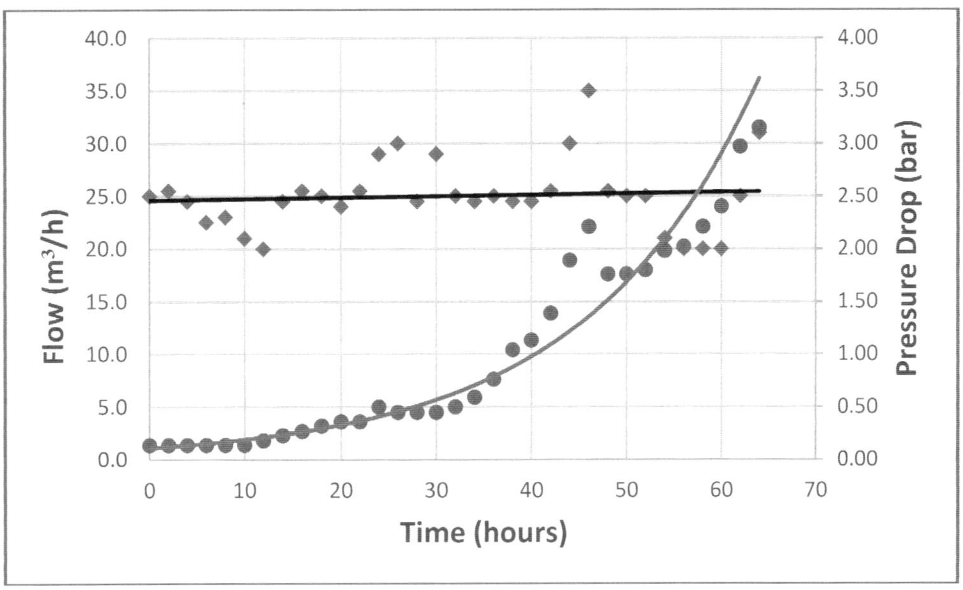

Conclusions

- A precoat of Ecosorb S-492 on the secondary filters significantly reduces the color feed to the existing GAC and the resulting decolorized liquor after the GAC, allowing the refiner to:
    - Handle much higher carbonated liquor color with no increase Fine Liquor or refined sugar color
    - Significantly reduce the fine liquor color feeding the crystallization process when processing "standard" carbonated liquor, resulting in either:
        - Produce the same quality refined sugar as before, but with increased refined sugar yields, by modifying the washing in the centrifuges
        - Producing lower color refined sugar, which may be of value in certain situations

- An Ecosorb precoat can be implemented in any refinery with polish filters prior to GAC or IER using existing equipment, with no adverse effect on production rates or daily refined sugar capacity. In fact, the increase in refined sugar yield will increase the daily refined sugar capacity.

Acknowledgements

The authors would like to thank the management and operators at the Taiwan Sugar Refinery referenced in this presentation for allowing us into its refinery, and for agreeing to let us present the data.

References

Delgado, I. and Lengen, J. (2016) NOVEL ADSORBENT FOR IMPROVED COLOR AND TURBIDITY REDUCTION WITH REDUCED PROCESS LOSSES. SIT paper #1130.

Abstract

In 2016, following a review of the refinery's off-site odour impact on proposed local developments, the Redpath Sugar Refinery in Toronto, Canada changed their carbonatation to a patented liquid CO_2 sourced process. This paper is a review of the decision making process that lead to this change, an overview of the equipment and processes involved and a summary of the experiences of the operation of the plant to-date.

Contents

Introduction

The Redpath sugar refinery opened in 1959 on a deserted waterfront property near downtown Toronto. Almost 60 years later the refinery is a 650,000 te/yr capacity plant refining sugar from Central and South America with carbonatation clarification. The deserted waterfront location is now the centre of one of the largest urban redevelopment projects currently underway in North America. Focus on efficient refining operations needs to be shared with maintaining the refinery's interests during the waterfront redevelopment process.

Background to the Process Change

Redpath sugar was built on reclaimed land approximately 1km from then downtown Toronto. Here is the site prior to the construction of the refinery

Site before construction

While the refinery grew gradually in output through the 1960's, 1970's and 1980's, the city of Toronto developed and the waterfront became a hub of industrial and commercial activity, centered on materials supplied from the waterways.

Redpath Sugar in Toronto's Industrial Port Setting

Over the next twenty years industry gradually moved away from Toronto harbour, leaving commercial and residential developments in its place. The result was the Redpath sugar refinery was the last industrial user of the harbour in downtown Toronto.

By 2013, a residential development was completed adjacent to the refinery and the plans for the development of the remaining lands continued at a fast pace. The result of this land development will be a refinery entirely neighboured by commercial and residential occupants.

Redpath Sugar is now amidst a rapidly evolving residential / commercial setting

A key consideration during this process is the off-site environmental impact of the refinery. New developments that potentially put the refinery out of future environmental compliance need to be identified and the potential issues addressed before planning permission is approved so the eventual occupation does not result in environmental complaints that cannot be economically addressed.

This work has been ongoing for over a decade. During the developments up to 2015, consideration was given to the common off site environmental parameters; noise, gaseous emissions and dust. However in 2015 a new environmental parameter was introduced by the Ministry of Environment and Climate Change, **odour**.

Potential Toronto Waterfront

Commentary regarding the smells associated with the refinery are not new to Redpath. A quick search on the internet will reveal the refinery's position in the "10 Most Notorious Smells in Toronto" list (burned marshmallow being the descriptor). However the introduction of odour as environmental parameter required a change in approach from the management team. The science of odour measurement was unfamiliar to the Redpath team and it is worth dedicating a few sentences to this.

The Science of Odour Measurement

Odour is measured in odour units (O.U. Dimensionless). Defined as,

Number of unit volumes of odourless gas required to dilute one unit volume of odorous gas (under standard conditions) to reach the odour panels detection threshold. The accepted reference value is equivalent to 123 μg n-butanol (CAS 71-36-3). Evaporated in 1 cubic metre of neutral gas (at standard conditions) this produces a concentration of 0.040 μmol/mol.

Odour Panel

Composed of 8 assessors who are qualified to judge samples of odorous gas using dynamic olfactometry. Each assessor must be appropriately screened and meet the selection criteria outlined in the EN 13725:2003 standard.

Detection threshold (Sample)

Dilution factor at which the sample has a 50% probability of being detected by a human assessor.

Although the regulatory system is based on investigation of complaints received, odour measurements were used by Redpath to identify likely sources of future complaints. In effect the benchmark was set such that there should be not more than 1 odour unit (O.U.) at each receptor i.e. no odours that can be detected by 50% of the population. Although residential balconies are not required to meet the requirements for noise, they *are* required to meet the requirements for odour.

Odour Measurement in Practice

Redpath engaged odour experts to measure the odour coming from the site. This began with a device known as the "Nasal Ranger Field Olfactometer", which looks like this:

The instrument has a selectable dilution ratio on the front to allow the user to measure the O.U. though dilution with air passing through a carbon filter. The nasal ranger was used to identify sources of odour from the site.

Redpath also used an odour panel for analysis. Sample balloons were filled with odour at full concentration from odour sources and were sent to the odour panel for concentration measurement and characterization. Beyond O.U. two other 'measurements' of odour were taken. The *Hedonic Tone (HT)* of the odour; a measure of the level of pleasantness or unpleasantness of the odour. HT ranges from -10 (least pleasant) to +10 (most pleasant) and is independent of the character of the odour. *Characterization* involves identifying the odour with pre-determined odour descriptors from an 'Odour Wheel'. See Appendix A.

After months of work on odour testing (and over C$500,000 spent) approximately 50 potential odour sources were investigated. The outcome of which was the carbonatation vent stacks were identified as the most problematic source of odour. The results of the testing from the stacks were:

Measurement	Result average / *(range across 3 stacks)*
Odour Units (O.U.)	16931 / *(11039 to 26217)*
Hedonic Tone (HT)	-1.7 / *(-10 to 6)*
Odour Character (See Wheel)	*Earthy, Offensive, Fishy, Floral*

The remaining odour emission sources were less concerning in O.U, Hedonic Tone and Character than the carbonatation stacks. Odour modeling was carried out to model concentration of the carbonatation odour at receptors at the proposed developments, some of which resulted in concentrations in excess of the limit of 1. The conclusion of the testing and modeling work was that complaints from new development receptors would be unlikely if the carbonatation vent stacks were eliminated.

Odour Mitigation Options

Several options for odour mitigation were considered. Scrubbing and a process change to phosphatation were considered. However elimination of the carbonatation vent stacks with a process change from boiler flue gas to a pure CO_2 supply was chosen as the change with the least associated process risk and given tight timetables for compliance, the certainty of eliminating the problem odour. The Redpath engineering team, with technical support from ASR Research and Technology developed the liquid CO_2 carbonatation plant design.

The Redpath Sugar Liquid CO_2 Carbonatation Plant

The heart of the liquid CO_2 carbonatation plant at Redpath consists of two new reactors and a liquid CO_2 storage and evaporation system.

Simplified PFD for Toronto Liquid CO_2 Carbonatation System*

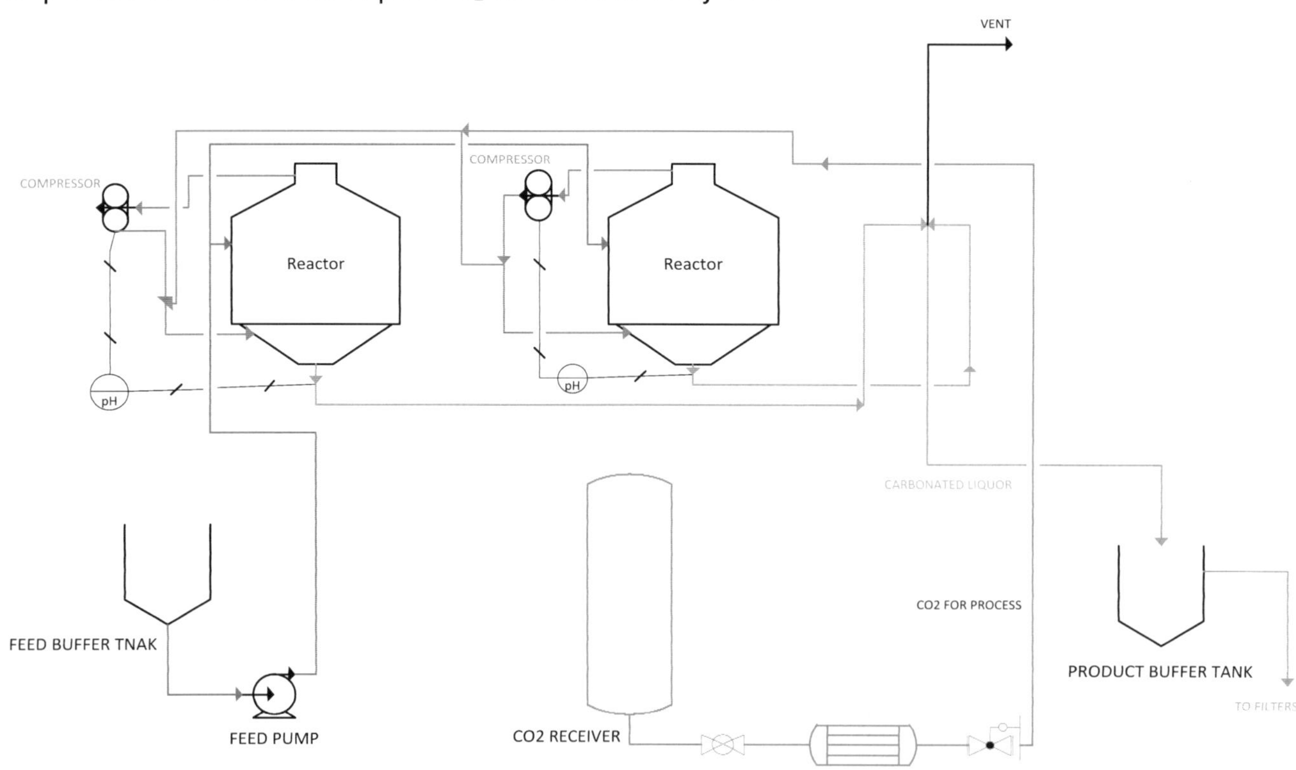

*Note: Patents Pending

Liquid CO_2 is supplied by 30 MT road tankers and is stored in a 50 MT tank. Control valves control CO_2 flow into the carbonatation reactors by maintaining a pressure in the reactor headspace, with the pressure in the storage tank maintained at 20 barg using an electric heater. CO_2 is supplied from evaporation in a water-heated evaporator using waste heat from the vacuum pan condenser water or from the top of the tank in gaseous form. The entire system from storage and evaporation to carbon dioxide piping and the reactors was all located outside of process buildings to minimize any risk of CO_2 leaks affecting oxygen levels in the factory.

The fresh CO_2 is introduced into the gas recirculation system of the reactors with the reaction rate controlled with a cascade control loop with pH of the discharging carbonatated liquor controlling the speed of blowers that recompress the gas from the reactor headspace to the gas distributors in the reactor. The original lime slaking system was retained.

Commissioning

Commissioning of the new carbonatation plant presented some challenges, mostly associated with the liquid CO_2. Redpath partnered with a CO_2 supplier, who designed and installed the CO_2 storage and evaporation system, replenishing the CO_2 inventory on telemetry. Early on, it became clear we had undersized this system. The refinery demand for CO_2 was higher than the original calculations resulting in exceeding the evaporative capacity of the CO_2 plant at times. Insufficient heating from the original electrical heater lead to liquid CO_2 passing unevaporated through the heater and expanding across the downstream gas pressure regulating valve. The consequences of this can be seen on examination of the phase diagram of carbon dioxide.

Carbon Dioxide Phase Diagram

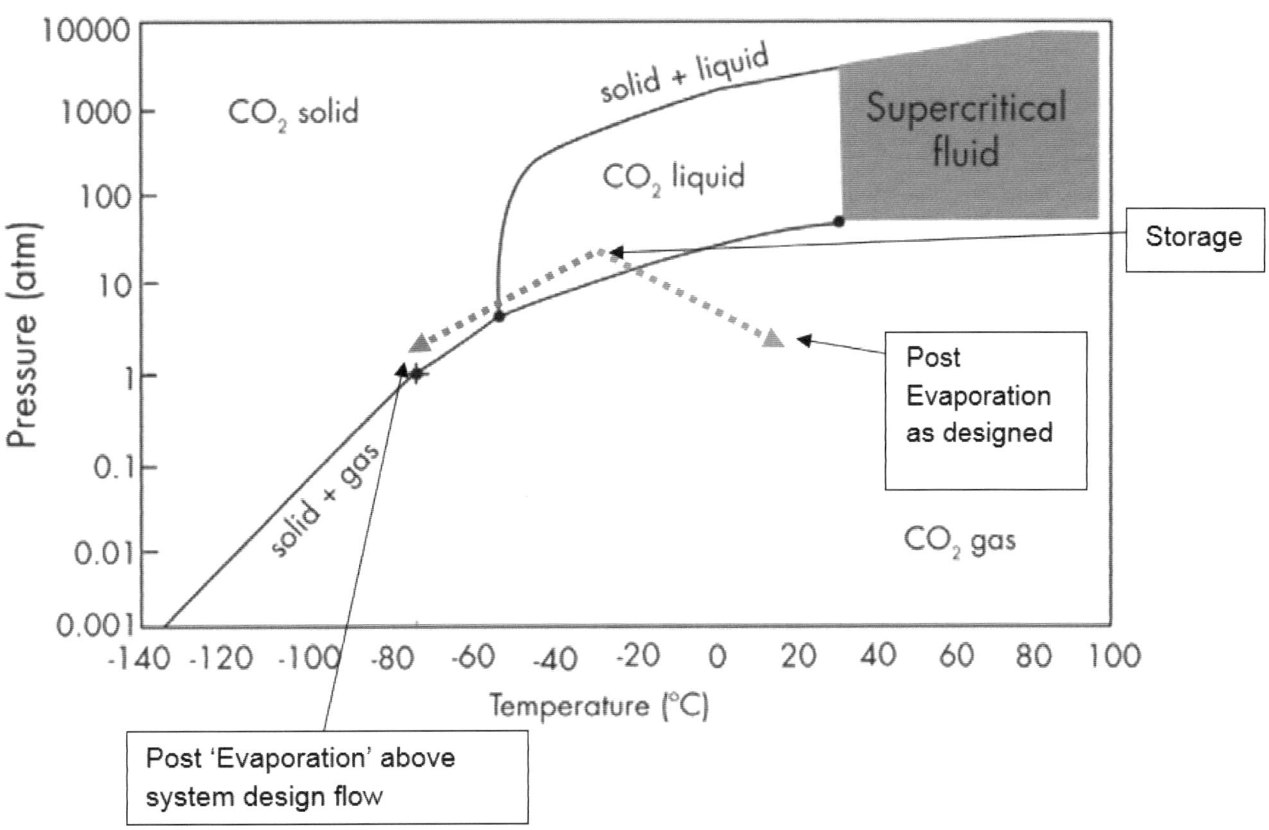

Attempted evaporation of liquid carbon dioxide without sufficient heat input results in migration to the 'solid' part of the phase diagram. In the piping system designed to transport gaseous CO_2 to the reactors, 'snow' formed and blocked the lines at temperatures below -60°C, resulting in plant shutdowns while the line thawed (quicker in the Canadian fall than the Canadian winter). In addition to the line blockage resulting from undersizing of the system, the original design did not include a low temperature cut out to detect this phenomenon and shut the system down before line blockage occurred. During the first few months of operation blockages in the liquid CO_2 evaporation system resulted in repeated production losses.

The CO_2 system was a turnkey installation with the equipment rented for a small annual fee as part of the supply contract. With the large volumes of CO_2 required for carbonatation, it became clear that Redpath Sugar was actually a very large user of CO_2 presenting engineering challenges that were not common in typical installations. On one occasion a relief valve on the evaporation system froze open.

As commissioning continued, the evaporative capacity of the plant was increased several different iterations of evaporator were used as the equipment required to evaporate the rate of CO_2 required was not readily available. The development of the evaporation system was as follows:

1. Original 60 kW evaporator. Solid aluminum with electric heater elements.
2. 18 kW ambient evaporator, added to supplement 60 kW evaporator.
3. Modification to take evaporated CO_2 from the top of the tank to utilize the tank heater as additional evaporation capacity.
4. Upgraded 100 kW unit installed as the new tank heater to increase the amount of gaseous CO_2 that could be taken from the tank headspace.
5. Water heated shell and tube evaporator used during plant operation, the heat source is the warm water from the pan condensers.

In operation, the CO_2 supplied from the tank headspace and the water heated exchanger differ significantly in temperature.

Evaporator	CO_2 inlet temperature (°C)	CO_2 outlet temperature before expansion (°C)	CO_2 outlet temperature after expansion (°C)
Electric tank heater	-22	-22	-40
Water heater	-22	26	15

A low temperature cut out is now installed which shuts the liquid outlet from the tank if the temperature downstream of the gas-regulating valve is <-10°C. This protects against the dry ice formation in the discharge pipe, however as the plant is normally started on the electric heater with the associated low temperature of the gas, the low temperature cut out needs to be overridden prior to starting the water heater.

Current design of the CO_2 evaporation plant.

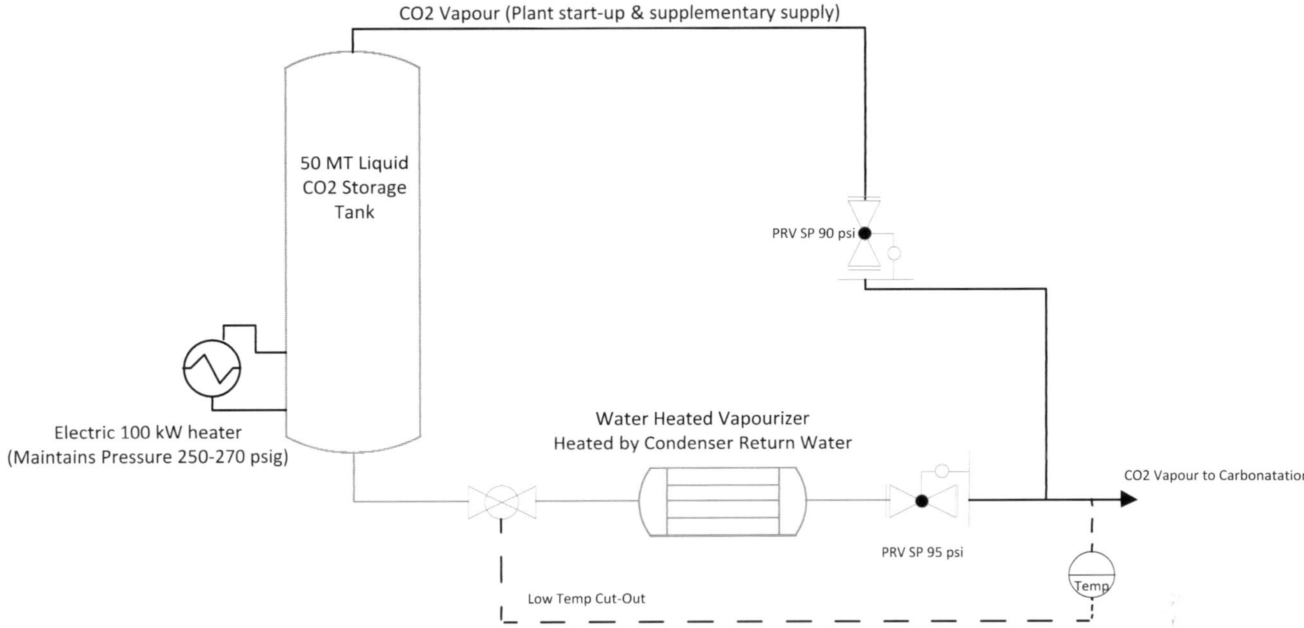

Carbonatation Operating Experience

Here are the key observations after 18 months of operating the new carbonatation plant:

Reaction Kinetics

In operation the new carbonatation reactors hold a significant inventory of CO_2 compared to the traditional design. This inventory of pure CO_2 in the headspace of the reactor, under pressure and undiluted with nitrogen and oxygen leads to very quick system responses to disturbances in lime or feed liquor flow, where the pH can drop by 2 units in 2-3 minutes. In the traditional carbonatation system, the inlet gas valves close in response to the reduction in liquor feed flow reducing the associated pH reduction. With pure CO_2 care must be taken to avoid fluctuations in lime flow and feed liquor rate.

Gas Distributor Cleaning

Before liquid CO_2, it was necessary to clean the gas inlet distributors twice per year. This required a confined space entry from a high-pressure water jetting contractor. There was also build up on the walls of the reactors that could come loose and cause restrictions in the discharge piping system. The experience with liquid CO_2 has been that there is no appreciable build-up and no associated need for cleaning of the inlet gas distributors.

Carbonatation Process Performance

The charts below show the colour removal and lime use on site. It should be noted, that the most accurate measurement of lime consumption is the weight of incoming lime deliveries. Although carbonatation is the largest lime user, other clarification and pH adjustment process that consume lime are included in the data. At Redpath the performance in filtration is used to set the lime ratio used in carbonatation, colour removal is not the primary driver.

Commissioning of liquid CO_2 was completed in July 2016. Inspection of the lime consumption trend versus the raw sugar starch trend suggests that filtration performance since the liquid CO_2 commissioning requires lower lime percentage on melt per unit starch than the traditional carbonatation. Colour removal, it could be argued has tended to be lower since the change to liquid CO_2. There are multiple factors in colour removal and filterability on a plant scale, so firm conclusions cannot be drawn from this data. Further work is required.

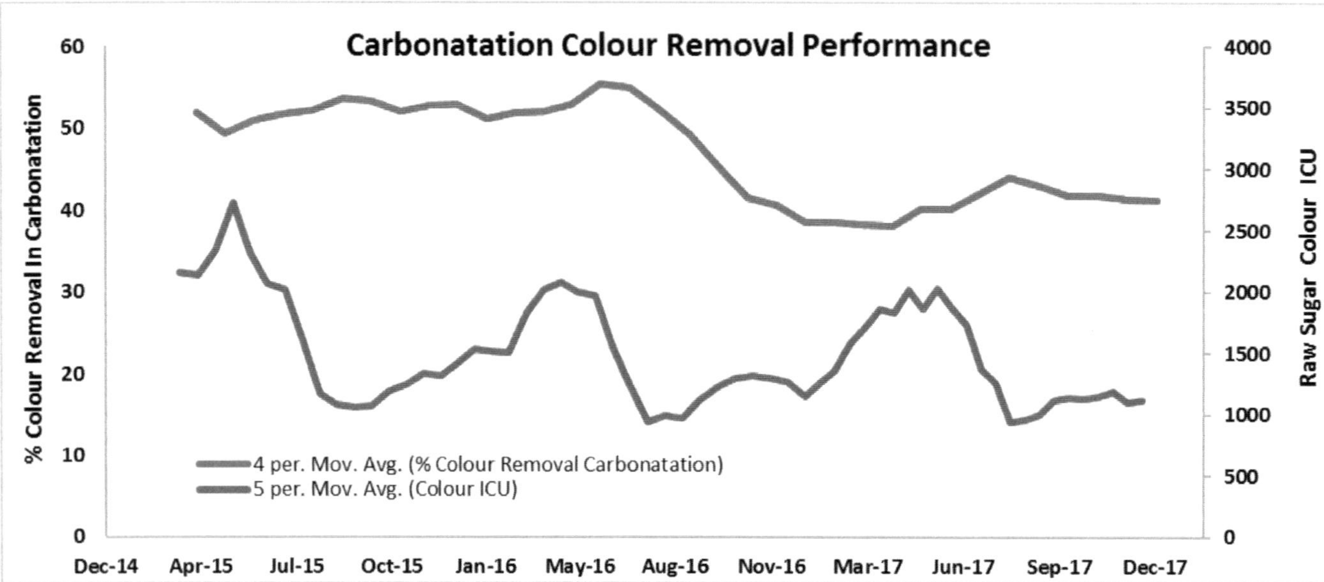

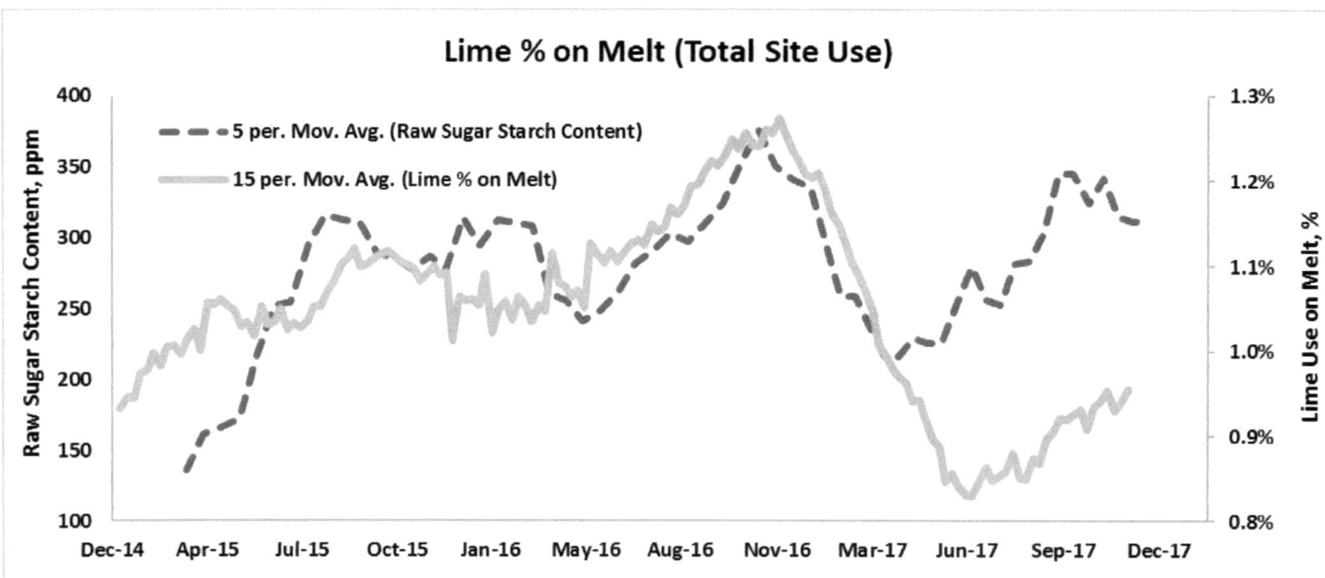

Cost Benefit Analysis

While the change in carbonatation process was driven by environmental performance, the return on investment was analyzed during the project design. Key considerations were:

1. *Purchase cost of CO_2.* Liquid CO_2 in Toronto is approximately C$100/MT, there is also a small rental charge for the liquid storage tank and evaporation system. The initial calculations for CO_2 consumption indicated a usage rate of under 0.5% lime on dry solids melted, actual consumption is closer to 0.6%.

2. *Reduced heat and water losses* from the carbonatation vent stack. The equivalent of 3 MT/hr of steam was estimated to be lost from the original carbonatation system. The energy equivalent of 4% of the steam consumption of the refinery was required to reheat the liquor from this heat loss. This heating was achieved with a combination of evaporator vapour heating in contact heaters with direct steam injection as back-up. The new carbonatation plant does not have any significant venting so this heat loss was eliminated in the project. The reheating steam used would be saved and the evaporator vapours that were being used could be redirected to replace other steam users, which were the affination syrup heater and the granulator/silo air heaters.

3. *Change in electrical load of the plant.* The NASH liquid ring flue gas compressors were not efficient users of electricity, so eliminating these from the plant should have been a good source of financial savings. The net electrical savings were promising at 430 kW; -650 kW from shutting down the NASH compressors, +220 kW in new consumers, liquor pumps and the recirculation blowers.

So 18 months on, how is the project performing from a financial perspective? As previously mentioned the initial CO_2 consumption was underestimated so there is an associated increase in the CO_2 costs along with some inflation. There was no temperature loss across the new reactors so both the direct steam injection and vapour for liquor heating were eliminated. Initially, there were no users for the newly available vapour, so there were no immediate savings – the vapour ended up in the evaporation tail condenser. The granulator air heaters and the affination syrup heater have now been commissioned, however further users for the vapour need to be connected to realize the full energy savings. Electrical savings reflect the fact that the full steam savings have yet to be achieved.

Project Costs/Savings Estimates Table	*Estimated* annual savings (costs) from project	18 Month Post project *estimate*
1. Purchase cost of liquid CO_2	(C$260,000)	(C330,000)
2. Natural gas savings from eliminating lost energy from carbonatation stack.	C$550,000	C$410,000
3. Net electrical savings from eliminating large compressors for boiler flue gas minus new drives in project.	C$60,000	C$150,000
Net	C$350,000	C$230,000

So a move to liquid CO_2-based carbonatation has a financial payback in Redpath's case. If it is ultimately proven that liquid CO_2 based carbonatation does impact colour removal and lime consumption, these parameters need to also be factored into the economic analysis. At Redpath, where lime consumption influences significant mud disposal costs in addition to lime and CO_2 purchase costs, any reductions in lime on melt results in additional financial savings.

Next Steps

Work is still in progress to realize the full benefits of the liquid CO_2 system. The plans in the short/medium term are:

1. Find a user for the surplus vapour. The likely consumer for this is the HP sweetwater supplied to the affination melter. At present this is not heated and enters the melter at low temperature. Circulating this through one of the spare contact heaters will improve the energy performance.
2. Fix the remaining issues with the liquid CO_2 evaporation system.
3. Modify the exhaust pressure of the main boiler to provide pressure to the high pressure first effect evaporator. This will increase the HP turbine to full load and eliminate site electrical import,

reducing this cost to zero and resolve the reduced payback in gas-saving projects that exists at present.

4. Reduce the lime and CO_2 consumption. A project team is working on a Lean Six Sigma project to identify the drivers of lime consumption with the aim of making a significant reduction.

Summary and Conclusions

So when taking into consideration all that has been learned in the last 18 months does Redpath recommend converting to liquid CO_2? Every site is unique, but if the site considering conversion has the following characteristics then a conversion could be advantageous.

- High cost electricity, or an electrical demand that needs to be reduced.
- Primary steam is used for carbonatation pre/post heating that would be directly saved if the heat loss from carbonatation is eliminated.
- A strong local market for liquid CO_2 with economic pricing and at least two suppliers.
- A site burning or considering burning biomass in its boilers which may contain fertilizers, or pesticides where the impact of the combustion products on food are undesirable or unknown.
- A site looking to get steam and power from an off-site cogen plant, where CO_2 compression is uneconomic.

Liquid CO_2 carbonatation does not come without its complications and the design of the CO_2 evaporation plant is critical to avoid its reliability becoming an unwanted distraction from sugar refining.

At Redpath, the carbonatation conversion to liquid CO_2 achieved its primary objective, which was to prevent a compliant refining operation being taken out of environmental compliance by the proposed developments around the refinery. The secondary objectives of an economic return look promising, further process development is still underway to maximize the potential of the technology.

When considering the bigger picture of an industrial facility gradually surrounded by commercial and residential development, the approach taken by the senior leadership team at Redpath is a great example. Working closely with a strong legal team and environmental and land planning experts from the early stages of the process has worked to preserve the refinery's interest to this point. As a result, the Redpath sugar refinery has a long future as part of the Toronto waterfront and serves as the model for industrial establishments facing similar challenges.

Acknowledgements

The author would like to acknowledge Balasahep Patil, Narayanan Seshadri and Jonathan Dunn for their assistance in the preparation of this paper. They have each spent many hours trying to defrost frozen CO_2 lines.

Appendix I – Odour Characterization Wheel

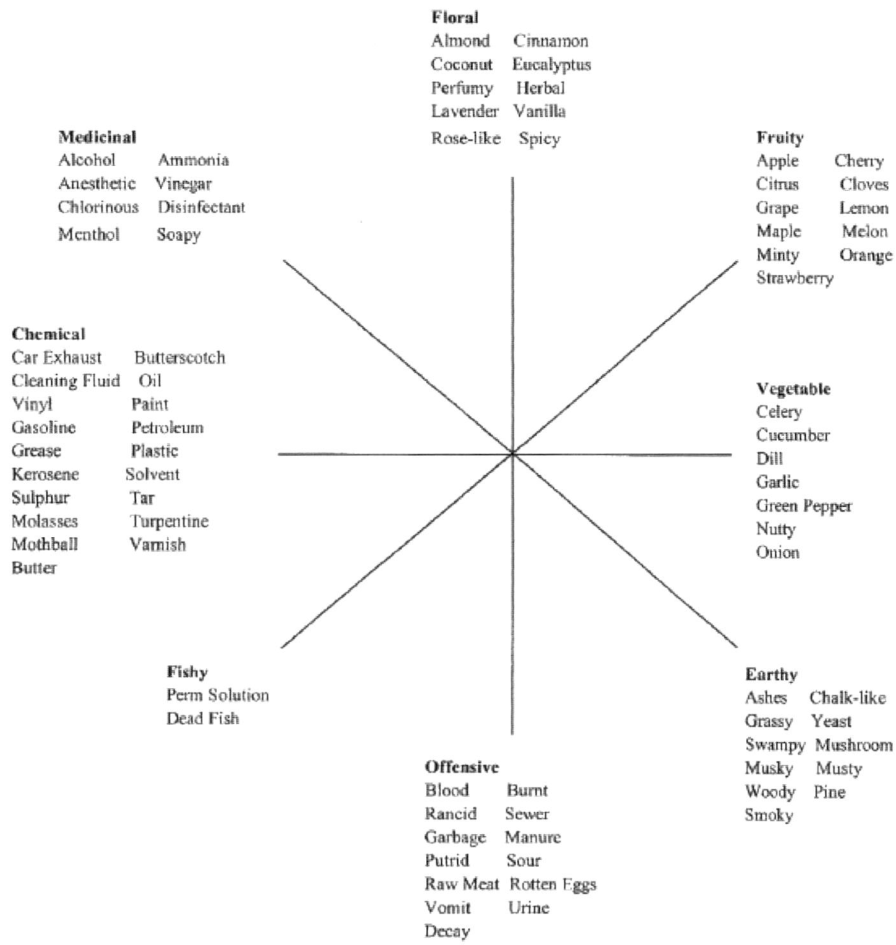

Efforts for reducing the Thermo-Acidophilic Bacilli levels to Japanese Major Company's requirement

Naoyuki Matsuda, Masumi Hayasaka and Kazuhiro Ootsuki

Quality Control Section, Fukuoka factory, Mitsui Sugar Co., Ltd.

11-30, 6-Chome, Hakozakifuto, Higashi-ku, Fukuoka-shi 812-0051, Japan

TEL+81-92-651-2031

E-mail: Naoyuki Matsuda@mitsui-sugar.co.jp

Introduction

The large-scale spoilage accident of apple juice that occurred in West Germany in the 1980s was due to the pathogenic bacteria called *Alicyclobacillus acidoterrestris* also abbreviated as AAT. Since this bacterium is a thermophilic bacterium that prefers acidity under aerobic conditions, it is classified as Thermo-Acidophilic Bacilli and also called TAB for short.

Thermo-Acidophilic Bacilli is a generic name for bacteria which as the name suggests are heat-resistant and develop favorably in acidic environments. In Japan as well, spore forming bacteria that grow in acidic beverages and produce pathogenic agents having an offensive odor have been reported from commercially available acidic beverage cans in 1990. TAB control became an important issue from the latter 1990s in quality control at beverage manufacturers for example that included measures such as adopting oxygen-permeable small plastic bottles for acidic beverages(1). As a sugar refining company which supplies sugar to beverage manufacturers, we could not afford to ignore taking measures to control TAB.

We also had a case in which it was pointed out that an offensive odor occurring in our customers' beverage products was due to the presence of TAB in granulated sugar, resulting in suspension of shipments and we were asked to take measures to prevent recurrence.

In this paper, we report on efforts to reduce TAB and to meet TAB level required by recent manufacturers.

1. What is Thermo-Acidophilic Bacilli (TAB)(1) (2)

① General characteristics

Certain microorganisms are known to grow under severe conditions and survive there for a long period of time. These are generally of a type that forms spores. These are also heat resistant, so they do not usually die even when heat sterilized at about 100°C. Bacteria that can grow even under acidic conditions are bacteria called TAB. The features are summarized below.

i. **Thermophilicity** : It grows at 55°C or higher, and proliferates better at such higher temperatures.

ii. **Aerobic property:** It cannot grow under an oxygen free environment.

iii. **Acidophilic property:** It grows only under acidic conditions.

iv. **Heat resistance:** Conditions of 106°C for 30 seconds or 120°C for 20 seconds or less are required for sterilization.

v. **Nutritional requirements:**

There are no specific nutritional requirements such as vitamins, amino acids, etc. It grows even in beverages with little fruit content or in "Near Water," which is a soft drink with minute amounts of flavoring.

② **Main types**

Although over a dozen varieties of Thermo-Acidophilic Bacilli have been found, not all of them are pathogenic bacteria having an offensive odor. Table 1 lists the characteristics of representative bacteria species.

The most problematic bacterium is acidoterrestris. A characteristic of this bacterium is that it grows well at relatively low temperatures compared to other TABs and produces guaiacol which is a pathogenic agent having an offensive odor.

Table 1. Typical TAB species

Species	Acidoterrestris	Acidocaldarius	Cycloheptanicus
Optimum growth pH	3.0	4.0	4.5
Growable pH	2.5–6.0	2.0–6.0	2.5–6.0
Optimum growth temperature (°C)	45	50	50
Growable temperature (°C)	25–60	35–70	30–55
Offensive odor component(Guaiacol)	(+)	(-)	(-)

2. Reason why measures were required at Mitsui Sugar

In the summer of 2007, a major manufacturer we will call Company A, that operates nationwide informed us of complaints of offensive odors occurring in their beverage product. Results from investigation identified the pathogenic bacterium as TAB, acidoterrestris. Further investigation of the raw materials revealed that pathogenic bacterium was found in granulated sugar from our Fukuoka Factory, so shipments were stopped. The reason why the abnormality occurred only in the product using sugar from our Fukuoka factory is that the product is manufactured and stored at normal temperatures in all regions, and since the Kyushu area where Fukuoka prefecture is located has a climate hotter than most other areas in Japan. It was inferred that the temperature was suitable for TAB growth especially in the summer.

Since we were advised that negative control is necessary to resume shipment, the Fukuoka Factory set up a countermeasure committee and started working on how to resume customer shipments.

3. Conventional TAB control level and target control level

The conventional delivery standards and the analysis values of the products having problems were as follows.

Delivery standard : 5 cfu/100 g or less (cultured at 50°C)
The product with the problem
: 0 to 2 cfu/100 g (cultured at 50°C)
35 cfu/15 kg (cultured at 50°C)

Although the analysis values of the products were within the product specifications, it was found that in the beverage product, in cases where even one acidoterrestris is present in the beverage and the growing conditions are satisfied, the bacteria will grow and develop an offensive odor. Therefore, for sugar to be manufactured from here onwards, we aim to control acidoterrestris by establishing a target with the following numerical values. Acidoterrestris is capable of surviving at low temperatures, so we started culturing it at 35°C as a test target.

Target	35°C cultivation	:	0 cfu/5 kg
	50°C cultivation	:	5 cfu/5 kg or less
	Guaiacol	:	Negative

4. Sugar refining production process at Mitsui Sugar

Figures 1 and 2 show the sugar refining production process of Mitsui Sugar Fukuoka Factory.

① Liquor end (LE) process

Figure 1 is a simplified diagram of the LE process. In the LE process, raw sugar is converted to magma in a mingler and the washed sugar is separated by an affination centrifugal. This is then melted in a melter to obtain raw liquor (RL). Next, the RL is subjected to carbonation, and brown liquor (BL) is obtained after filter filtration. The BL passes through bone char to become char-liquor (CL), then passes through an ion exchange resin to obtain an ionic liquor (IL), and passes through a ceramic filter to remove microorganisms to obtain a fine liquor (FL). After this, the liquor is sterilized with a UV sterilizer and is concentrated by an effective evaporator.

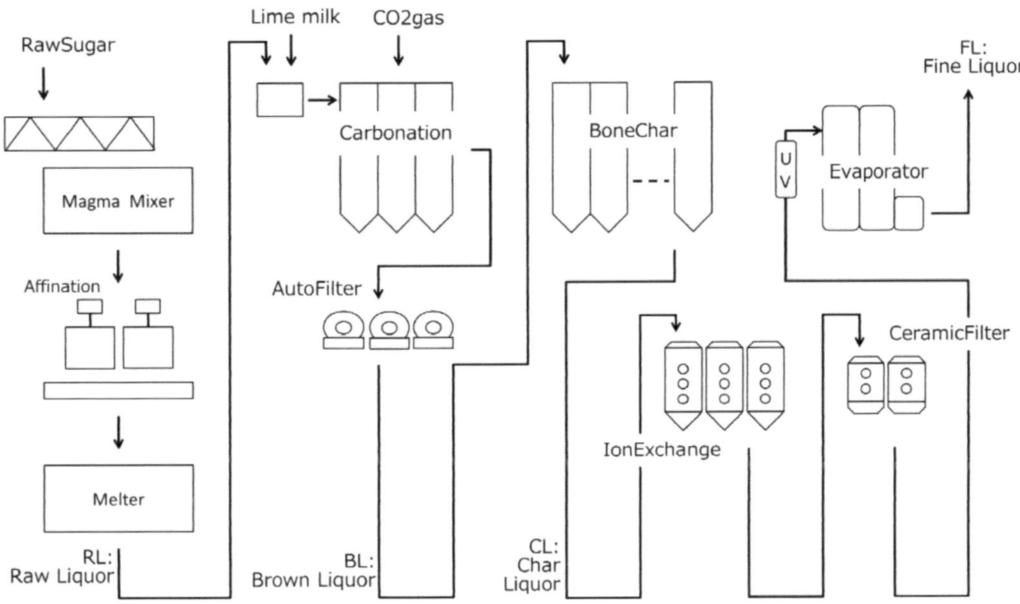

Figure 1. LE process chart

② Sugar end (SE) process

Figure 2 is a simplified diagram of the SE process. The FL is crystallized by a vacuum pan, and granulated sugar is obtained by separating massecuite. The granulated sugar is dried with a dryer, a fluidized bed cooler cools it and removes the powder. It is then stored in a sugar bin. Sugar comes out from the sugar bin is then sorted with a vibration sifter. It is packed in each packaging form after passing through magnets and metal detectors.

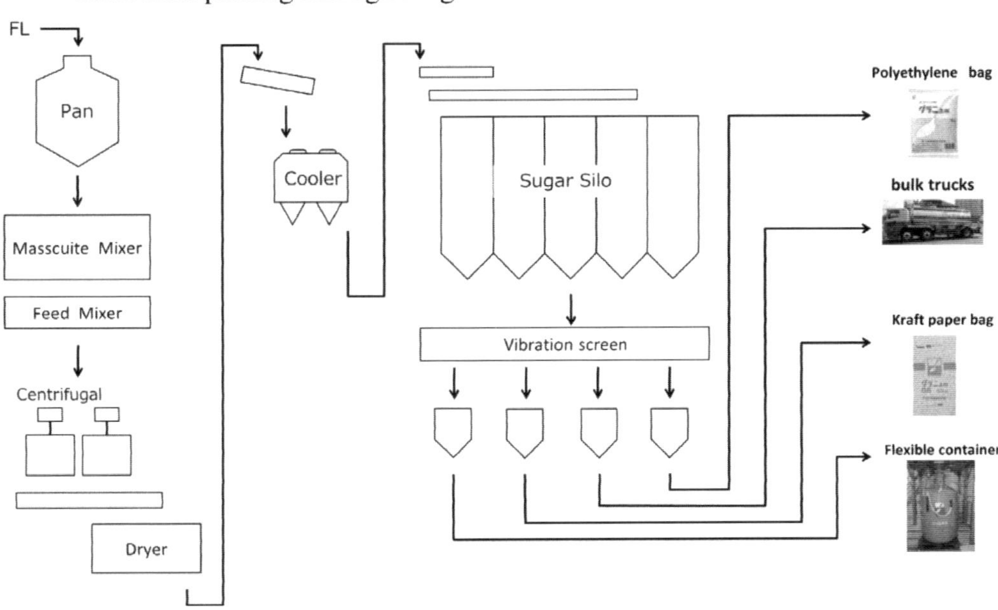

Figure 2. SE process chart

5. Efforts to achieve a TAB zero count

We set a goal of "zero bacteria" for the product and set up a countermeasure committee to achieve this goal.

① Review of check points and countermeasures

We created a checklist that identifies the steps and work tasks for producing granulated sugar and checks whether there are points where countermeasures need to be taken or points needing improvement. Based on this, we investigated check points and countermeasures.

② Checks of TAB occurrence points

Next, based on the checks in ①, we checked the status of TAB occurrence in the factory. As a result, Table 2 shows points where many TAB occurrences were detected. Since bacteria were found in the parking lot, the raw sugar warehouse, and the manufacturing floor, there was concern this TAB was being brought in from the outside environment. In addition, although the ceramic filter is used by pre-coating the filter-aid (diatomaceous earth) onto the filter material, TAB was present inside the tank which mix the liquor and filter-aid for pre-coating. In addition to this, since TAB was also present inside the FL tank, it was reconfirmed that elimination and sterilization of bacteria in the LE process are essential. Furthermore, since bacteria are present in the product centrifugal, it was found that thorough washing and sterilization are necessary to prevent proliferation in the SE process.

Table 2. Points where many TAB are detected

	TAB detection amount (cfu/10 cm^2 wiping inspection)	
Check points	35°C culturing	50°C culturing
Parking lot	4 cfu/1 g	4 cfu/1 g
Raw sugar warehouse floor	2	2
Vacuum Pan floor	17	0
Ceramic filter Inside a tank of filter-aid for pre-coating	15	194
Inside a FL tank	4	300
Inside a product centrifugal	0	71

At the same time, we conducted checks of raw sugar that fukuoka factory usually use and obtained the results in Table 3. Since TAB was detected in culture media at 35°C in all production areas, it was found that TAB exists in raw sugar.

Table 3. TAB amount in raw sugar

Production area	35°C culturing (cfu/50 g)	50°C culturing (cfu/50 g)
Australia	11	0
Fiji	124	4
Thailand	3	0
Kagoshima (Japan)	1	0

Table 4 shows the results from checks of liquor and products. In these checks, many TABs were detected in CL and IL. Generally, it is believed that most of the bacteria die because of the highly alkaline state during carbonation. However, it was suspected that TAB grows in the bone char process because it was detected in CL and IL. Yet it was not detected from FL and products. This fact suggests that TAB can be removed and sterilized by improving and modifying the manufacturing process.

Table 4. TAB amounts in liquor and product

Process	35°C culturing (cfu/450 g)	50°C culturing (cfu/450 g)
BL	-	5
CL	-	176
IL	35	19
FL (pre-UV)	7	1
FL (post-UV)	0	0
Product	0	0

③ Measures

Based on the results from the checks, we decided to carry out process remodeling and facility improvement by way of three measures: to prevent bringing in bacteria; to prevent bacterial growth; and to thoroughly eliminate and sterilize bacteria. The actual measures taken separately for each process and environment are summarized below.

I. Countermeasures by changing sterilizing and process control methods

● Countermeasures in LE process

- **Higher sterilization level measures at start of production**

  We decided to sterilize all lines from the post-carbonation filtering machine down to the FL tank by using sodium hypochlorite at the start of production. In addition, the bone char equipment at the Fukuoka Factory is sweetening offed in the tower after Liquor passage, and the whole bone char is then withdrawn and sent to the kiln for reburning. By sterilizing the inner wall of the empty column with sodium hypochlorite, TAB growth in the bone char equipment was prevented.

- **Change melter temperature setting from 65°C to 70°C**

  Since acidoterrestris grows at a relatively low temperature among TABs, by raising the control temperature in the melter and setting a high alkaline condition in the carbonation column of the next step, we were able to boost the sterilization ratio.

- **Steam sterilization washing of the precoat tank**

  At the ceramic filter, the filter element is pre-coated with a filter-aid (diatomaceous earth) before passing liquor. By installing new steam piping onto the tank for mixing liquor and filter-aid for pre-coating and then performing steam sterilization before and after use, TAB growth was prevented.

- **Change filter aid used in ceramic filter**

  We employed a filter-aid having lower permeability than the conventional filter-aid (Table 5) to ensure reliable elimination of bacteria.

Table 5. Features of filter-aid for evaluation

Type	Average particle size (μm)	Permeability	Re-precoating frequency
Conventional product	23.9	0.50	10 to 50 days
Modified product	20.9	0.15	About 7 days

● **Countermeasures in SE process**

- **Higher sterilizing level before production begins**

  During sterilizing measures before the start of operation, it was decided to also spray sodium hypochlorite and alcohol onto the conveyor already cleaned with hot water to sterilize the TAB.

- **Extension of steam sterilization time after boiling in vacuum pan**

  After boiling in vacuum pan, the inside of the pan is normally steam-washed. However, we extended this steam washing time from 1,600 to 3,300 seconds, so any bacteria that might exist will definitely be sterilized

- **Intense sterilization of product centrifugal**

  Before centrifuge the products to be delivered to Company A, we decided to steam-sterilize the inside of the centrifugal basket and perform hot water washing of the conveyor under the centrifugal.

- **Increase the volume of wash water at the centrifugal**

  We increased the amount of wash water used at product centrifugal by 30% and completely isolated the syrup on the crystal surface, thereby reducing the possibility of TAB remaining in the product.

II. Expansion of UV sterilizer

Conventionally, the UV sterilizer was installed in the front process of the evaporator. Since there were cases where slight TAB growth was confirmed in evaporator, UV sterilizer having the same performance was added to the post-process of the evaporator, so that TAB can be fully sterilized.

III. Expansion of ceramic filter

Removing almost all TAB was now possible by changing the filter-aid for the ceramic filter, but in some cases these could not be completely removed depending on the pre-coat condition, so a ceramic filter was added and by connecting it in series with a conventional filter, we aimed at completely removing all TAB.

IV. Measures concerning the environment

- **Paving of parking lots**

  TAB is a soil-derived bacterium. The parking lot at the Fukuoka Factory was unpaved, so we took the precaution of fully paving the entire parking lot.

- **Installing a shoe sole washer**

  In order to eliminate the risk of bringing TAB adhering to shoe soles onto the premises, we enforced zoning control on the premises and installed shoe sole washers at all entrances to the hall.

- **Overall cleaning**

  The inside of the factory was cleaned by all employees and sterilized by alcohol spraying where necessary. In addition, we have set fixed dates for carrying out cleaning of the entire facility on a regular basis from this time onward.

6. Result of efforts

Results of the above countermeasures are shown in Table 6. This table shows the transition in TAB quantity within FL and products in the order of measures taken. In the product itself, the TAB quantity was reduced to the target level and we achieved zero detection of acidoterrestris bacterium which produces the unwanted guaiacol.

Table 6. TAB transition in 5 kg of FL and product

Order of implementation	Measure	FL (cfu/5 kg)		Product (cfu/5 kg)	
		35°C	50°C	35°C	50°C
1	Before measure	up to 10	up to 15	ave. 2	ave. 12
2	Thorough sterilization	up to 5	up to 15	0 to 3	ave. 7
3	Expansion of UV sterilizer	0	0	0 to 1	ave. 5
4	Expansion of ceramic filter	0	0	0	ave. 4 {AAT (-)}

After these efforts, we submitted a product sample to Company A and passed their inspection (bacteria negative in guaiacol production). In addition, after being inspected by Company A, our factory and process were rated as "No problem" and shipments were resumed.

7. TAB level required by major customers

Table 7 shows the TAB level required by major customers currently dealing with Mitsui Sugar.

Each customer is a manufacturer that manufactures foods such as beverages and jellies and their products all contain a large amount of water. Although few customers request submission of samples for inspection before delivery, they do require establishing special specifications.

Table 7. Examples of special anti-TAB specifications

Customer	Item	Individual specifications	Test frequency	Cultivation temperature	Cultivation period
Company A	TAB	Less than 5 cfu/100 g	Per lot	50 ± 2°C	5 days
ditto	TAB	Less than 5 cfu/100 g	Per lot	36°C	5 days
ditto	Acido-terrestris	Negative	At time of TAB detection	45°C	-
Company B	TAB	0 cfu/10 g	Once a month	45°C	5 days
Company C	TAB	0 cfu/10 g	Per lot	45°C	5 days
Company D	TAB	1,000 cfu/50 g	Per lot	45°C	5 days
ditto	Acido-terrestris	Negative/50 g	At time of TAB detection	45°C	-
Company E	TAB	0 cfu/10 g	Per lot	45°C	5 days
Company F	TAB	Negative/10 g	Per lot	45°C	5 days

8. Summary

One comment that Company A gave us after implementing the current countermeasures was that, "Mitsui Sugar's Fukuoka Factory has the best sugar quality in Japan."

The Mitsui Sugar Fukuoka Factory was able to produce quality sugar that satisfies their standards sufficiently even in conventional process control, and it was more than comparable to the quality of other companies in the same industry in Japan. However, in order to increase customer satisfaction, we will always have to pursue quality above original.

Implementing the current countermeasures has proven that TAB is in fact controllable.

The Mitsui Sugar Fukuoka Factory will continue to maintain its operation at zero AAT and continue its efforts to become a model factory for TAB control.

One big benefit we reaped from this experience is that we completely reformed our employee's situational awareness by taking this opportunity, which in a way is even better than our success in improving the manufacturing process.

References

(1) *The Unified Test Method for Thermo-Acidophilic Bacilli*, Japan Fruit Juice Association

(2) *Thermo-Acidophilic Bacilli—Alicyclobacillus Bacteria—*, Supervised: Akira Yokota, Tateo Fujii

APPROACHING ZERO EFFLUENTS FOR ION EXCHANGE DECOLORIZATION SYSTEMS IN SUGAR REFINERIES

Marc André Théoleyre, Anne Gonin, and Dominique Paillat,
EURODIA Industrie, ZAC St Martin, Impasse St Martin, 84120 Pertuis, France

Presentation Abstract

In 1998, the first unit designed to reduce effluents from decolorization resins system was presented: this was achieved thanks to a nanofiltration membrane unit. Different improvements have been proposed over time to decrease the consumption of salt and water, and to limit the environmental footprint of such a process. Twenty years later, thanks to a further combination of membrane technologies, we are able to provide a system to recycle close to the totality of the salt and water required during standard regeneration operations. With this process, only a minor part of the resins depollution stream would have to be discharged, once every 10 to 15 cycles.

The proposed process is based on the properties of various types of membranes: (1) nanofiltration is effective to achieve a good separation between salt and color bodies, but leads to a slightly diluted stream compared to the standard brine concentration needed for the regeneration; (2) Reverse osmosis is effective to recover water from a diluted brine but leads to an intermediate retentate concentration stream, too low for reuse; (3) Only electrodialysis allows the transfer of salt from a stream with an intermediate brine concentration (the reverse osmosis retentate), to a higher concentrated stream (the nanofiltration permeate), thus allowing to reach the required concentration for resin regeneration. The salt consumption is therefore limited to the portion exchanged with the resin. Water consumption is only limited to the final concentrated retentate, mixed with the molasses, plus the water needed for the membrane cleaning operations.

Such a process is currently being installed in a new sugar refinery under construction in the Middle East

1- Introduction

All over the world, the limitation of the environmental impact of operations is a constant concern in the sugar industry. Generally speaking, the reduction of effluents must lead to the decrease of water consumption by recycling it as much as possible and by separating organics from minerals in order to enable recycling of those components in secondary factory streams, molasses, brine, etc.

In sugar refineries, membrane technologies have demonstrated their efficiency in separating organic and mineral compounds from effluent streams in order to recover and eventually recycle them. The first system that was developed in the early 2000's was Nanofiltration (NF) to recover and recycle up to 75% of the brine needed for resin regeneration. This system has been then widely installed in new refineries where Ion exchange resins are used for decolorization.

Since this first experience, many trials and research have been conducted for membrane selection (regarding cutoff, pH or temperature resistance...), allowing for continuous improvements in efficiency and performance for effluents production and salt recovery.

Recently, EURODIA has realized that, even with improved performance, there were cases where a 75% salt recovery was not enough to meet the environmental requirements in terms of waste disposal but also in terms of water usage. This has been the driving force for EURODIA to develop new processes.

2- Process principles

Ion exchange resins are widely used in refineries for sugar decolorization. Color bodies, mainly consisting of macromolecules with weak acid functions, are fixed to anionic resins in the chloride form. Two exchange mechanisms have been identified: adsorption on the hydrophobic resin matrix thanks to hydrophobic bonds and ion exchange between the weak organic acid function of the colorant and the chloride ions of the resin. Regeneration is achieved by flowing an alkaline brine at relatively high concentration. The high pH of the brine (12 to 13) promotes the ionization of the colorants and their solubilization. Brine concentration (at 100 g/l) decreases the hydrophobic interactions and, due to resin shrinking (an effect of osmotic pressure), leads to the desorption of the colorants fixed by the resins. Studies have shown that regeneration efficiency is severely reduced with lower concentrations, even when the brine volume is increased.

This regeneration step takes place through a resin bed full of water and, followed by a rinsing step, leads to the dilution of the effluent. Figure n°1 shows a classical profile representative of the evolution of effluent concentrations during the regeneration steps. Table n°2 presents the compositions of the different fractions identified during the regeneration steps.

Fig n°1: Regeneration profile of resins used for sugar decolorization.

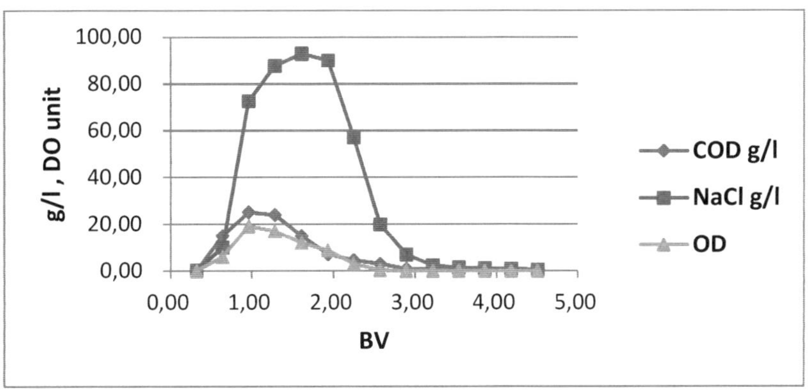

Table n°2 : Composition of eluates and mass balance..

Regeneration level : 150 g salt per liter of resin.

Fractions	BV	Nacl g/l	DCO g/l	NaCl g/l resin	% of Total Salt
High concentration fraction	1,60	80.41	15.23	128.66	85.77
Low concentration fraction	1.30	9.72	4.86	12.64	8.43
Recycled water	1.60	0.84	0.27	1.35	0.90
TOTAL	4.50	31.70	6.91	142.65	95.10

The dyes present in that effluent are mainly high molecular weight aromatic components that are known to be very difficult to degrade, especially since they are in high salt concentration solutions. Salt removal is mandatory before sending such a solution to a biological water treatment station.

The regeneration mass balance shows that only 5 to 10% of the chloride ions from the regeneration brine are exchanged during regeneration. The major part of the required salt is not consumed. It is recyclable, provided that one can separate it from the dyes and adjust its chloride concentration.

Three fractions are identified, the first (1.6 BV) contains the higher salt concentration; it is reusable for regeneration after color removal and salt concentration adjustment. The second (1.3 BV) is very diluted and cannot be reused without prior concentration. The third, corresponding to the first effluent produced and the final rinsing effluent, is reusable as is for brine displacement. Due to molecular weight difference between salt and color bodies, the separation of these two species is relatively easy by using nanofiltration membranes and has allowed this application to become a standard in sugar refineries since the beginning of the 2000's.

Reverse osmosis is the most cost-effective way to recover water from a salt stream; it is the reference technology for potable water production from sea water. However, the use of the technology is limited because of the osmotic pressure of the concentrated retentate. In industrial economic conditions, this pressure is around 50 bars, corresponding to a 50 g/l salt concentration in the retentate, which is much too low for reuse as regeneration stream.

When a maximum of salt must be recovered, a third technology is needed in order to adjust the concentration of the recovered brine: evaporation or electrodialysis. If evaporation is well known in the sugar industry, most sugar people are not familiar with the second. Electrodialysis is a membrane-based technology in which ions are moved by applying an electrical field. It is based on the use of ion-exchange membranes placed in an electrical field. Ion-exchange membranes are dense membranes, having a chemical composition similar to ion exchange resins. They are constituted of a copolymer Styrene Divinyl Benzene matrix on which functional groups are grafted in order to obtain Cation-exchange membranes and Anion-exchange membranes. Such a functionality gives selective permeability properties to the membranes. Cation-exchange membranes are permeable to cations but not to anions, and Anion-exchange membranes are permeable to anions but not to cations.

The elementary Electrodialysis cell (see Fig. N°3) consists in a pair of anion and cation exchange membranes. An industrial process is obtained by a multiplication of cells organized in a "Stack". Thus, ions are displaced from one compartment to the next one where they are concentrated.

Fig N°3 The principle of Electrodialysis

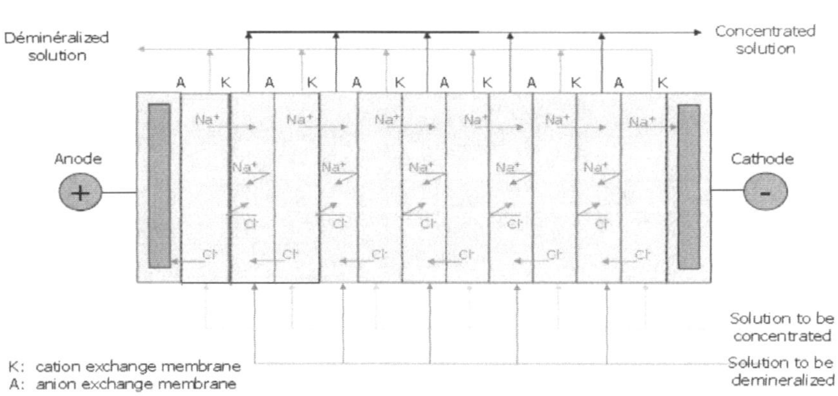

Electrodialysis is a mature technology; it was the original method for salt production in Japan from the beginning of the 50's where its economic reliability for salt concentration was demonstrated from sea water concentration up to 200 g/l before salt crystallization. It is perfectly well-suited to the concentration of a salty stream from 50 g/l to 100 g/l.

3- Industrial implementation and process development

In most of the situations (such as the European conditions), the goal is to recover the salt of the high concentrated fraction in order to decrease the salt concentration in the effluent sent to the waste water station. In some cases, the retentate is desalted by the way of diafiltration and concentrated as much as possible, then mixed with the refinery molasses. The permeate volume, 1.4 to 1.5 BV, has a salt concentration around 80 g/l . Adjusting its concentration to 100 g/l by adding concentrated brine leads to the production of a perfectly adapted stream, both in terms of volume and concentration, to resin regeneration. It corresponds to the recycling of 75 - 80% of the salt needed for resin regeneration. In this case, the low salt concentration stream is considered as an effluent.

Fig n°4: Basic Nanofiltration process for salt recovery

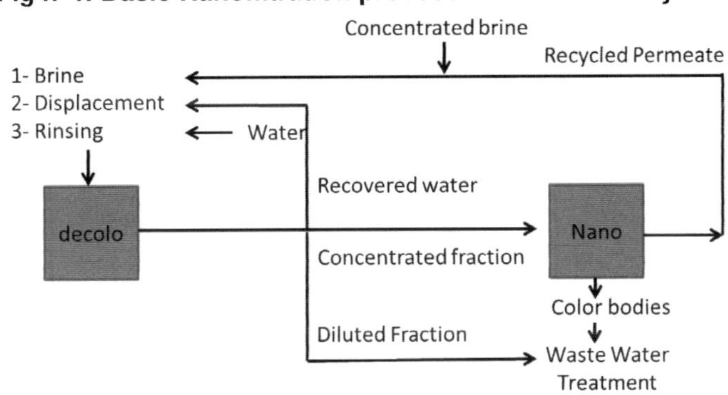

In some highly regulated situations, as in the Middle East, water recycling and waste water volume limitation are very strict constraints and these impose to also recycle the salt of the dilute fractions. The zero-effluent target requires the treatment of the totality of the salt streams. However, the volume, 2.9 BV, and the concentration, 55 g/l, of the overall effluent are not adjustable by adding concentrated brine and a concentration step is mandatory.

The regeneration efficiency requires the use of a 100 g/l brine. Due to the osmotic pressure of such a concentration it should be necessary to have reverse osmosis unit working at up to 100 bars. Such a pressure does not seem feasible in economic industrial conditions. In Bahreïn, EURODIA has successfully achieved such results by recovering the totality of the effluent, treating it by nanofiltration and concentrating the recovered permeate by adding an evaporation step. This system has been proved to reduce drastically the amount of waste and water demand.

Fig n°5: Coupling Nanofiltration/Evaporation.

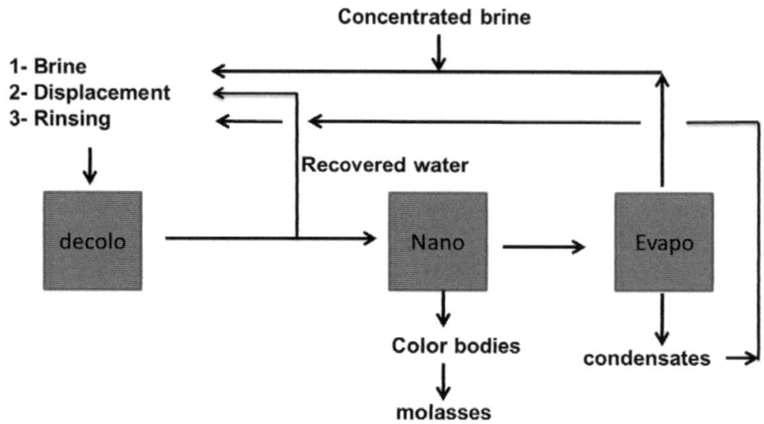

Based on these satisfactory results and performance, EURODIA has continued its efforts to improve the system, considering that the evaporation of brine streams is costly (due to specific construction materials) and probably not the best option in terms of energy requirements. Based on more than 25 years of experience in Electrodialysis in various fields of applications as an alternative to evaporation, EURODIA has developed a new patented process consisting of a combination of Reverse Osmosis and Electrodialysis in order to recover the salt of the diluted fraction. First, the high concentration fraction is treated through Nanofiltration membranes to obtain a decolorized permeate at a 80 g/l salt concentration. Then, thanks to Electrodialysis the salt content of the low concentration fraction (20 g/l), is transferred to the NF permeate and its concentration is increased from 80 g/l to 90 - 100 g/l. Nearly all the salt contained in the eluates is so recovered and the salt consumption of the decolorization unit is limited to the chlorides exchanged in the resins. In order to improve the electrodialysis electrical yield, a Reverse Osmosis loop is added to maintain the optimal concentration in the dilute fraction of the Electrodialysis to produce good quality process water.

Fig n°6: Coupling of Nanofiltration/Electrodialysis/reverse osmosis

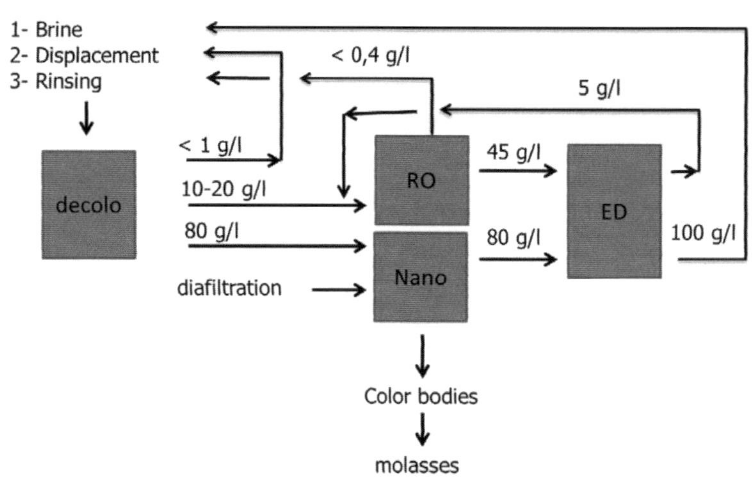

In Table 7, the performance of these three processes has been compared for a 1000 t/day sugar refinery, based on the EURODIA industrial experience, as well as to a conventional process without any recycling.

Table n°7: Comparisons – from no recycling to complete process with NF - RO – Concentration
Sugar production capacity: 1000 t/d

		Reference	NF	NF+EVAPO	NF+ED+RO
Consumables					
Salt	kg/D	6750	1375	350	350
Water	m3/D	285	135	65	65
Electricity	kwh/D		285	675	2650
Steam	t/D			18	
Effluents and coproducts					
Effluents	m3/D	219	75	19	24
COD effluent	kg/D	1300	750	7	7
NF Retentate	m3/D		7	4	4

When the effluent treatment is implemented, the Nanofiltration retentate can be mixed with the refinery molasses under the following conditions: the retentate volume and concentration must be such as the dry substance of the molasses is kept above 75% Bx. This means that diafiltration is needed to decrease the osmotic pressure of the retentate and to concentrate the retentate to the highest possible concentration, 20 – 25 % DS.

In the most complete processes, the final COD is limited to the effluents from membrane cleaning, or approx. 10 kg/day for a 1000 t/day refinery capacity. The water consumption is limited to the volume needed for the sweetening off. The operating costs correspond to the electrical and/or steam consumption of the installed equipment. The CAPEX is quite equivalent for both alternatives: Evaporation or Electrodialysis. The final selection has to be done based on the local costs for steam and electricity.

4- Conclusions

In sugar refineries, membrane technologies have demonstrated their efficiency in separating organic and mineral compounds from effluent streams in order to recover and eventually recycle them.

A Nanofiltration process was developed in the early 2000's to recover and recycle up to 75% of the brine needed for resin regeneration and this system has been then widely installed in new refineries using Ion exchange resins for decolorization. Since this first experience, many trials and research have been conducted for membrane selection (regarding cutoff, pH or temperature resistance...), allowing for continuous efficiency and performance improvements in terms of effluents production and salt recovery.

However, even with improved performance, there were cases where a 75% salt recovery was not enough to meet the environmental requirements in terms of waste disposal but also in terms of water usage and EURODIA has been very active to imagine, realize, implement and successfully start and operate the first refinery with a decolorization system combining resins + NF+ EVAPORATION. Based on this satisfactory results and performance, EURODIA has continued its efforts to improve the system and has patented an original process for brine/water recovery by combining NF + RO + ED, thus avoiding the need for Evaporation. This process will be soon implemented in a new sugar refinery in the middle east.

Based on its industrial experience, EURODA Industrie is able to propose the best combination of membrane technologies in order to meet the needs of its customers.

Bibliography :

(1) Botton H.R.; Use of bipolar membranes for ion exchange resin regenerant production. J. Chem.Tech. Biotechnol., 54, 341-347 (1992)

(2) Horbez D.; L'électro-électrodialyse et le traitement des effluents liquides. Informations Chimie n° 351 (1993)

(3) Cartier S., Théoleyre M.A., Decloux M.; Membrane technology in the sugar industry. SPRI meeting (1998)

(4) Cartier S., Théoleyre M.A., Decloux M.; Treatment of sugar decolorizing resin regeneration waste using nanofiltration. Desalination 113, 7-17 (1997)

(5) Verhaeghe F., Malgoyre R., Théoleyre M.A., Cartier S.; Industrial scale brine recovery by nanofiltration from ion-exchange decolorization. SIT (1998)

(6) Bento L., Machado Santos J., Hervé D.; Traitement intégré des éluats de régénération des unités de décoloration en raffinerie de sucre. IAA (1989)

(7) Bento L.; Sugar decolorization by ion exchange resins with regenerant recovery. Int. Sugar Journal, 92, 116-125 (1990)

Efforts to Reduce the Colour Load of Raw Sugar Entering a Sugar Refinery

Emmanuel M Sarir and Benhur Pabon

Carbo-solutions International, sarire@carbo-solutions.com

Keywords: VHP sugar, High Performance Decolorants (HPD), colour precipitants, clarification, crystallization

Abstract

Factory-scale trials were conducted at several mills with back end refineries to test the ability of high performance de-colorants (HPD) to decrease the colour of clear juice and raw syrup, and ultimately the colour of VHP sugar. Up to 30% colour removal was observed under ideal conditions. Centrifugal wash time was also reduced by 25%. Such performance would be interesting for factories focused on reducing the color load to their back end refineries or for factories wishing to avoid colour penalties for VHP sugar sold to a destination sugar refinery.

1. Introduction

The direct production of high quality raw sugar has a large economic advantage. Increasingly with green cane harvesting on the rise around the world, and factories having to process poor quality sugarcane at times, invariably results in the use of additional processes like syrup sulphitation, longer washing cycles in centrifugals, and remelting of final crystallized sugar with techno-economic disadvantages. As a step toward this goal, many studies have been undertaken to develop a simple and economical decolorization process to remove colored impurities in cane syrups. Technologies studied have included: double curing of C massecuite (Polanco et al. 2014), membrane filtration of juice, ion exchange and oxidative decolorants like ozone (O_3) and hydrogen peroxide (H_2O_2) (Getaz & Bachen, 1989; Davis, 2001), but the majority of these technologies have not been successfully adopted. This has been due to high capital investment and/or operational demands (i.e., high energy consumption and high maintenance cost).

1.1. Color and Color Removal

Color in sugar is a very complex mixture of organic compounds extracted from cane and formed during different process stages. Thus, color compounds can be classified into two main groups: natural colorants form sugarcane and those developed during processing. The majority of sugar colorants are anionic hydrophobic compounds with a wide molecular weight range. Their removal or elimination depends mainly on: the nature and molecular weight, the charge type and density (highly anionic, weakly anionic, amphoteric or weakly cationic), degree of hydrophobicity, and pH (Davis, 2001). The choice of a suitable de-colorization system, however, also requires an evaluation of all the parameters which include capital cost, operational cost, color removal effectiveness, process flexibility, and environmental impact.

1.2. Juice and Syrup Sulphitation

Sulphur dioxide (SO_2) has been used for centuries to produce plantation white sugars in many countries. The process is carried out by injecting SO_2 into the raw juice to a level of about 400 ppm SO_2. Factories using juice and syrup sulphitation have reported 25% less color in syrup and raw sugars. However, there are many problems associated with the use of sulfur: high sucrose inversion, increase in evaporators scaling, corrosion in pipes and process equipment and, in particular, environmental and health concerns. US-FDA (Food and Drugs Administration of the United States Government) currently has a 10 ppm limit on residual sulphur dioxide allowed in food products due to carcinogenic concerns. Furthermore, as an oxidative decolorant agent, sulphitation is hardly effective in eliminating caramels and glucose and fructose degradation colorants (Davis, 2001).

1.3. Syrup Clarification

In clarification of syrup by air flotation, very fine impurities are removed. The process involves primary flocculation with phosphoric acid and lime, polymer addition to increase size of flocs and physical capture of the micronized air bubbles within the flocs. This process reduces viscosity of syrup with benefits at the boiling stage. Although syrup clarification removes 80-90% of turbidity in raw syrup, color removal is low at only 5-10% and, therefore, necessitates the addition of decolorant polymers to increase decolorization (Rein, 1987). Significant colour removal is obtained onlt with high phosphoric acid and lime consumption.

1.4. Alternative Factory Boiling Procedures

Sucrose crystallization in vacuum pans can be performed using different boiling schemes. The choice of boiling procedure is determined by syrup purity and the sugar quality required. Most common boiling schemes are: two boiling system, three boiling system, double magma, and Very High Pol (VHP) scheme. VHP produces the lowest sugar color.

Currently, many raw sugar factories are adopting double magma or VHP boiling schemes in order to improve the quality of their sugar. Color reduction, however, sometimes is not high enough and process modification involves high capital investment. The boiling process is energy demanding and evaluation of overall sugar recovery must be undertaken to evaluate the effect of boiling scheme modifications (Madho & Davis, 2008).

1.5. High Performance Decolorants (HPD) and Color Precipitants

Both High Performance Decolorants (HPDs) and color precipitants are developed by Carbo Solutions International. They are both very similar in chemical composition and are engineered processing aids that can be added to select industrial processes to remove color. HPD are typically added to mixed juice for color removal or to inhibit the formation of color, replace juice sulfitation, and enhance juice clarification; they are particularly applicable for plantation white or very high pol (VHP; 99.0 to 99.49 pol) sugar production or syrup production.

Color precipitants are polymers of highly charged cationic polymer are designed to neutralize the negatively charge colouring matter in juice and syrups. These are applied to juices, syrups and even directly into massecuites boiling inside pans. , and their intended application is to syrups, or directly into massecuites in vacuum pans to avoid the inclusion of colorants in sugar crystals. The liquid precipitant contains active areas of strong adsorbent radical groups that can destabilize and trap impurities leading to their eventual removal. Since color precipitants are very low in viscosity and high in water solubility, they are completely and easily removed during the centrifugation stage so that they reside in the molasses not in the raw sugar. Thus, color precipitants help to improve the efficiency of the centrifugation process, and less washing time is required.

This paper discusses laboratory and industrial case studies on the use of HPDs and color precipitants to remove color and allow for the manufacture of high quality sugars in South and Central America. Traditionally, South American factories produce plantation white sugar using a sulfitation process, juice purification (clarification) with lime, double B and C magma boiling systems, with variations from country to country and from one industrial plant to another.

2. Experimental Section

2.1. Chemicals and products

HPD and color precipitants were provided by Carbo Solutions International LLC (Los Angeles, CA, USA).

2.2. *Process description*

The process involved the addition of HPD to mixed juice and color precipitants to syrup. This allowed an overall color removal of 25-30% enough to produce VHP sugar and even plantation white depending on dosage rate, process operational control and washing time at centrifugals.

2.3. Juice decolorization

HPD was dosed continuously to mixed juice tank after juice liming. Dosages varied from 20-40 ppm basis on cane crushed. Because the typical dose was between 20 to 40 ppm, for a sugar factory with milling capacity of 5,000 tons of cane per day (TCD):

HPD dosage: 5,0000 TCS x 20 ppm = 100 kg per day of HPD

3. Results and Discussion

3.1. Uses of High Performance Decolorants (HPD)

3.1.1. Juice decolorization

A comparative, 1 year study at a Colombian sugar factory (Table 1) showed that it is possible to eliminate juice sulphitation through the use of HPD, effectively reducing up to 0.05% more color in the cane syrup.

Table 1. Juice sulphitation replacement at Providencia sugar factory Colombia 2012 (Sarir et al. 2012). Average results listed.*

	Mixed Juice Color (IU)	Clarified Juice Color (UI)	Raw Syrup Color (IU)	Clarified Syrup Color (IU)	Juice Color Removal %	Syrup Color Removal %	Dosage (kg/MT cane)
Sulphitation	17,004	14,149	15,350	14,258	16.57	6.18	0.103
HPD	16,907	14,005	14,556	13,632	17.12	6.23	0.013

* HPD was dosed continuously to mixed juice tank after juice liming. Dosage varied from 20-40 ppm basis on cane crushed.

Similar results as in the Colombian sugar mill (Table 1) were observed at other Colombian and Ecuadorian sugar factories, when HPD products were tested in the facory laboratories. Example results from the Ecuadoran factory are shown in Table 2 and Fig. 1.

Table 2. Results from the application of HPD (MAJ-1^{TM}) at Ingenio La Troncal sugar factory, Ecuador in 2015. Average results are listed.

	Mixed Juice (A)	Clarified Juice (Blank)	Mixed Juice + HPD	Mixed Juice + HPD
		(B)	20 ppm	40 ppm
			(C)	(D)
Purity (%)	**79.3**	**80.6**	**80.4**	**80.8**
pH	**5.83**	**7.39**	**6.6**	**6.51**
Color (IU)	**9,914**	**7,321**	**6,420**	**6,232**
Turbidty (IU)	**34,463**	**2,826**	**2,969**	**2,857**
Conductivity mS/cm	**4.1**	**3.97**	**3.92**	**4**
Color Removal %		**26.2**	**35.2**	**37.1**
Turbidity (%) (%9Removal %		**91.8**	**91.4**	**91.71**

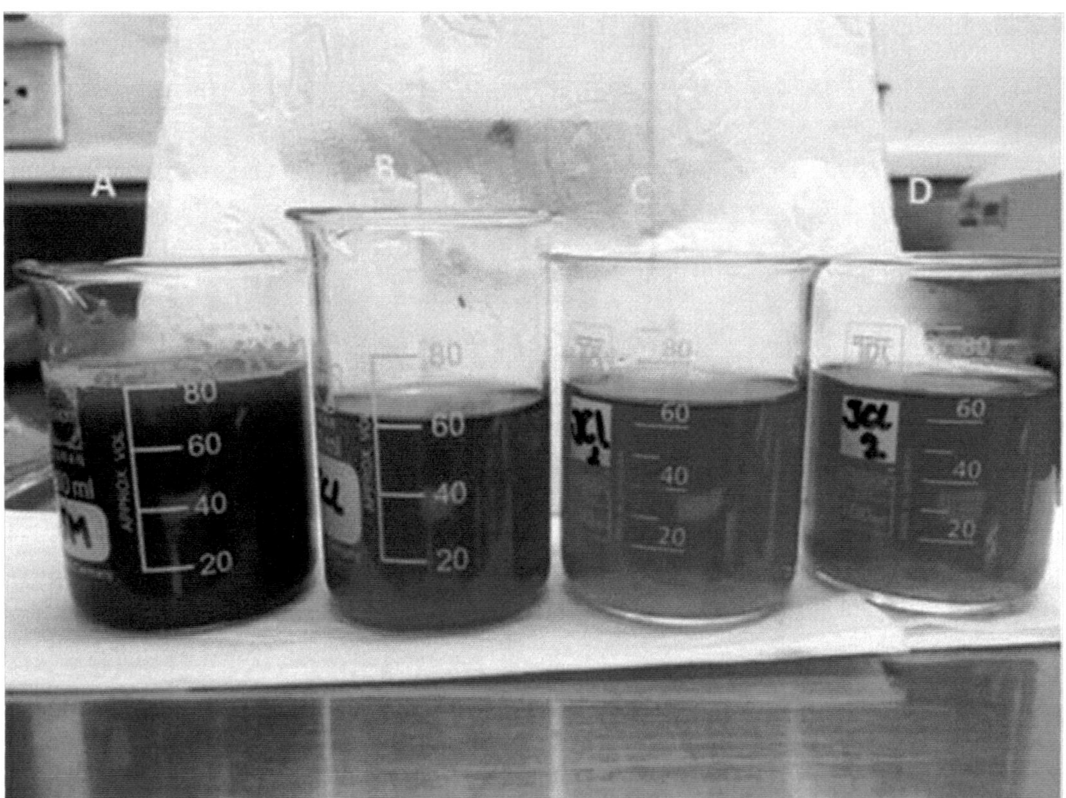

Fig. 1. Photographs to show the effectiveness of using HPD (MAJ-1™) at Ingenio La Troncal sugar factory, Ecuador in 2015. A is Mixed Juice, B is clarified juice, C is clarified juice plus 20 ppm HPD, and D is clarified juice plus 40 ppm HPD.

3.2. Use of Color Precipitants: Case Studies

Color precipitants can be added directly to raw syrups, or A and B massecuites in vacuum pans. This is because they are liquid products of low viscosity and, therefore, can be easily added to both syrup and massecuites using conventional metering pumps, as illustrated in Fig. 2.

The following case studies were undertaken at factories in South America, and highlight their practical use.

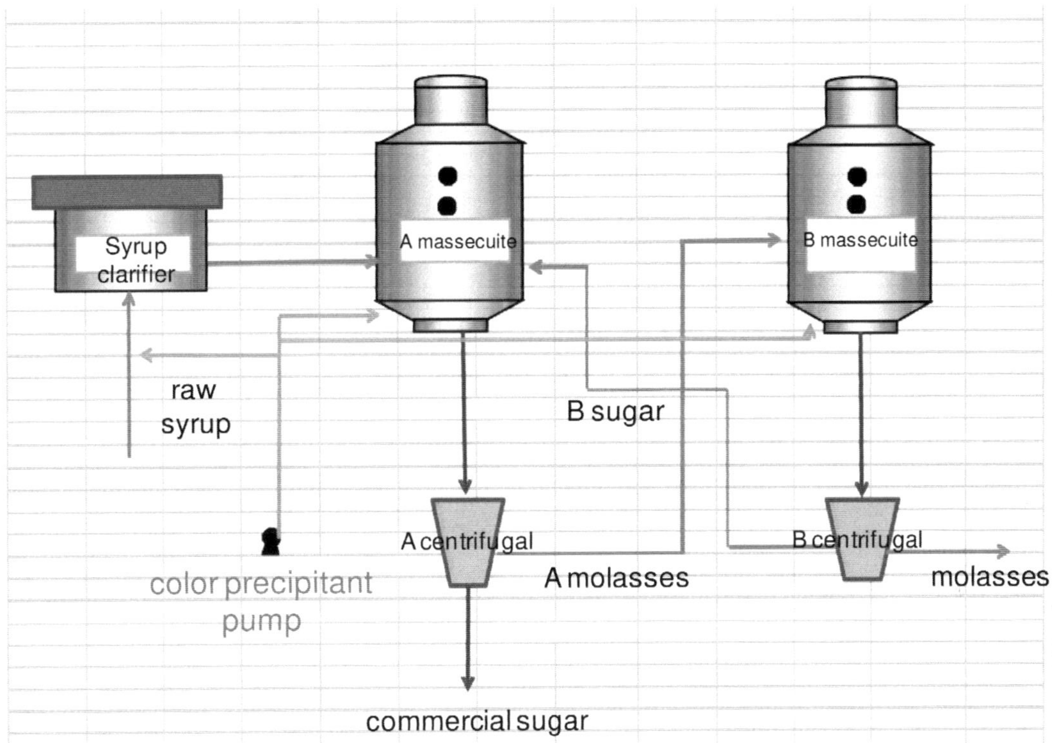

Fig. 2. Diagram to show the application of color precipant to syrup or massecuites.

3.2.1. Case study No. 1: Syrup decolorant addition at a sugarcane factory in Brazil

Color precipitant products were tested on the syrup clarification process at a sugarcane factory in Brazil and compared to the conventional process employed. The factory was looking to upgrade its process to (i) allow for the stable production of sugar Type 3A (ICUMSA color lower than 180 IU), and (ii) increase their sugar production based on better factory recovery (higher exhaustibility of final molasses).

Specifically, during the 2014/2015 processing season, the sugar factory tested the color precipitant CUA ST5TM. The precipitant was supplied via a metering pump, from a portable tank and applied directly into syrup after the last evaporator. The dose was kept constant in 260 ppm basis on produced sugar (Fig. 3).

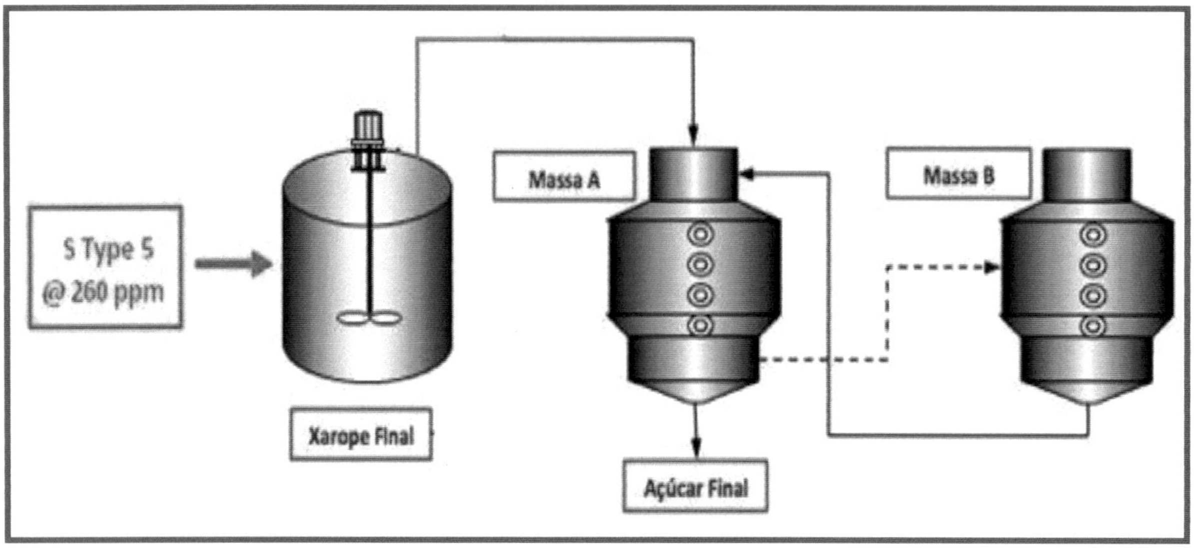

Fig. 3. Application of CUA ST5^{TM} decolorant (color precipitant) to the final evaporator syrup in a Brazilian sugarcane factory.

Results from the study (Case Study No. 1) are listed in Table 3.

Table 3. Color precipitant test at a Brazilian sugarcane factory in 2014. The color precipitant was CUA ST5^{TM}.

Factory Parameters	Traditional Process	Addition of CUA ST5^{TM} Color Precipitant
Color precipitant dose (ppm)	n/a*	260
Sugar color (IU)	165	148
Centrifugal wash time (sec)	16.5	12.5
Overall factory sugar recovery	74.6	78.5
50 kg sugar bags/ton cane	1.3	1.4
Final molasses per 50 kg sugar bag	41	33
Final molasses purity (%)	58.6	57.4
B sugar color (IU)	2,881	2,792

*n/a=not applicable

Overall, results in Table 3, showed that most factory parameters studied were improved by adding the CUA ST5^{TM} precipitant compared to the traditional process.

3.2.2. Case Study No. 2: Syrup decolorant addition at a sugarcane factory (San Carlos) in Columbia, South America

San Carlos sugarcane factory in Columbia did not utilize syrup sulphitation or vacuum pan (massecuite) oxidizers to control color but instead returned refinery liquor to syrup to carry on their special white sugar requirement. Unfortunately, when the produced sugar color was more than 150 IU, it was sold as C-type sugar at a loss. As an alternative to this existing process, CUA ST5™ color precipitant was studied by adding it at a dose of 400 ppm to A massecuites, 50 ppm to B massecuites, and 100 ppm on raw syrup on a sucrose basis. Results are listed in Table 4, and demonstrated that its use allowed for the direct production of special plantation white sugar less than 150 IU without any need of back boiling from the refinery.

Table 4. Color precipitant test at a Brazilian sugarcane factory in 2014.

Factory Parameters	Traditional Process	Addition of CUA ST5™ Color Precipitant
Color precipitant dose (ppm)	n/a*	550
Sugar color (IU)	200	122
Centrifugal wash time (sec)	12	8
Commercial yield (ton sugar/ton cane)	11.08	11.09
A type sugar (% total production)	12	67
Steam reduction (%)	n/a	0.1
B sugar color (IU)	4,962	3,538
B sugar turbidity (IU)	797	517

*n/a=not applicable

3.2.3. Case Study No. 3: Decolorant addition to massecuites at a sugarcane factory in the Dominican Republic

The use of color precipitant CUA ST5™ as a decolorization processing aid was also studied at a sugarcane factory located in the Dominican Republic, Central America. In this case study No. 3, CUA ST5™ was added directly to massecuites A and B at the Dominican Republic sugar factory. The color precipitant was added at the 50% massecuite level, at two doses of 100 to 200 ppm basis sugar and compared to the traditional process (0 ppm color precipitant).

Table 5. Color precipitant (CUA ST5TM) test at a sugar factory in the Dominican Republic.

	A Massecuite color (IU)	Commercial Raw sugar color (IU)	B Massecuite color (IU)	B sugar color	% Reduction in sugar color
Traditional Process	**14,376**	**1,452**	**34,426**	**14,647**	n/a
Color precipitant - 100 ppm dose	**15,809**	**1,001**	**12,026**	**7,405**	**31.1**
Color precipitant - 200 ppm dose	**23,147**	**1,323**	**33,321**	**11,190**	**8.9**

Findings indicated that despite increases in massecuite color with CUA precipitant treatment, there was markedly higher color removal in the final commercial A-raw sugar and intermediate B-sugar. This strongly indicated that there was a significant drop in color transfer from massecuites to crystal (see Table 5 and Fig. 4). As it is known that cane derived colorants transfer more readily into sugar than process derived colorants, the CUA must have chemically transformed or physically removed cane derived colorants preferentially. Furthermore, factory results were best at the lower 100 ppm dose of CUA, which is equivalent to significantly more cost savings.

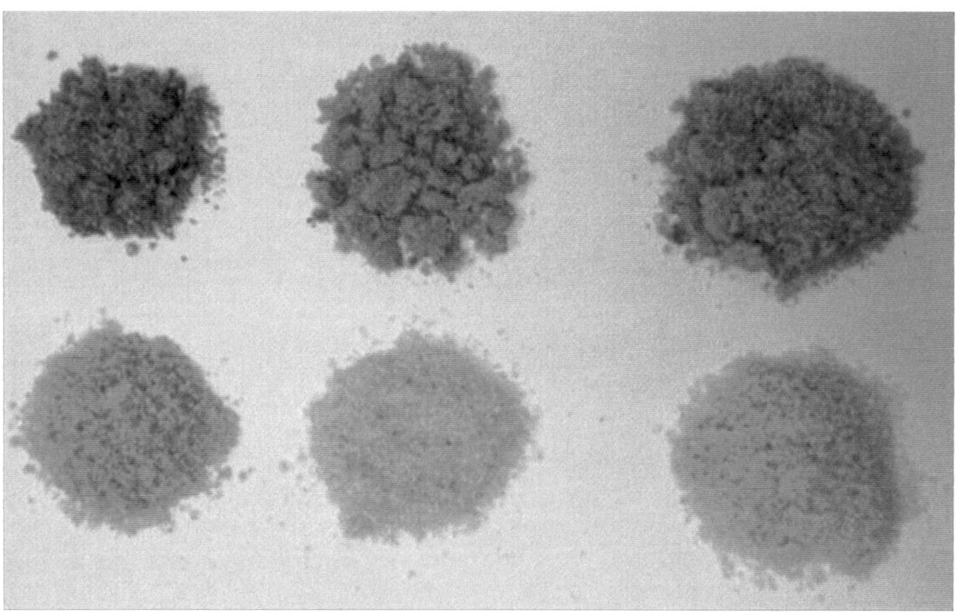

Fig. 4. Samples of A, B, and C sugars manufactured by the traditional process (top) compared to treatment with 100 ppm CUA ST5^{TM} color precipitant (below).

4. Conclusions

In three factory case studies the use of HPDs applied to mixed juice or color precipitants to raw syrup and/or massecuites in vacuum pans have been shown to enhance performance in the sugar process as follows:

- Up to 30% color reduction from mixed to clear (clarified) juice, which allowed the production of higher quality raws

- Eliminating syrup sulphitation to produce plantation white sugar
- Increased yield of sucrose as seen by shorter washing times in centrifugals (up 25%).
- Improved the color transfer rate at the pan boiling stage, to obtain 25-30% sugar color reduction compared to the conventional process

Overall, the use of HPD products and color precipitants offer significant advantages over traditional juice and syrup sulphitation color removal processes:

- HPDs and color precipitants are effective, easy to use, and safe to apply
- HPD and color precipitant use results in the reduction of lime consumption
- No sulphur residues occur in sugar and all the other benefits of a Sulphur free process
- Environmental friendly

5. References

1. Polanco LS; Kochergin V; Stephen Savoie S; Carline G. **2014.** Improvements of raw sugar quality using double purge of C-magma, Audubon Sugar Institute – Louisianan State University.
2. Getaz MA; Bachan, L. **1989.** Some ideas of the use of chemicals methods for improving color of A-sugar. *Proc. S. Afr. Sugar Technol. Assoc.*, *63*, 76-79,
3. Davis SB. **2001**. The chemistry of colour removal: A processing perspective. *Proc. S. Afr. Sugar Technol. Assoc., 75*, 328-331.
4. Rein P. **1987.** Syrup clarification in raw sugar mills. *Proc. S. Afr. Sugar Technol. Assoc., 61*, 22-31
5. Madho S; Davis SB. 2008. Review of proven technologies available for the reduction of raw sugar color. *Proc. S. Afr. Sugar Technol. Assoc. 81*, 165-183
6. Sarir EM; Jiménez Mazuera AM; Jaramillo Mora L; Botero, G. **2012.** Alternativaspara la optimización del proceso de producción de azúcar en fabrica, sustitución de la sulfatación de jugo, *Boletín Tecnicaña*, Colombia.
7. dos Santos R. **2015.** Efecto de la transferencia de color en la performance de la fábrica de azúcar. estudio de caso en usina Brasilera. Case study at a Brazilian plant.
8. Marín F, León W, Jimenez A, Jaramillo L, Donado C, Pabón B, Sarir E Add date Producción de azúcar blanco directo medianteaplicación de adsorbentedecolorante de alto rendimiento en meladuracruda y masas decocimiento Add publication & pages
9. Jimenez A; Cedano C. **2014.** Evaluation of ST5 color precipitant at La Romana sugar mill, Dominican Republic. *CSI internal document.*

International Sugar Journal
Agribusiness intelligence | informa

Policy makers and political masters

arvind.chudasama@informa.com

SternEnzym
The Enzyme Designer

Outline

Introduction – global sugar landscape

- Global sugar production
- Trends in global consumption
  - Per capita consumption
- Key drivers of supply and demand

Policy and policy making

Country case studies of policy making

- Proactive – EU
- Reactive - Brazil
- Political – India
- Apathetic – Caribbean Sugar Industry

Concluding thoughts

SternEnzym – Your Enzyme Designer to Improve the Quality of Your Sugar

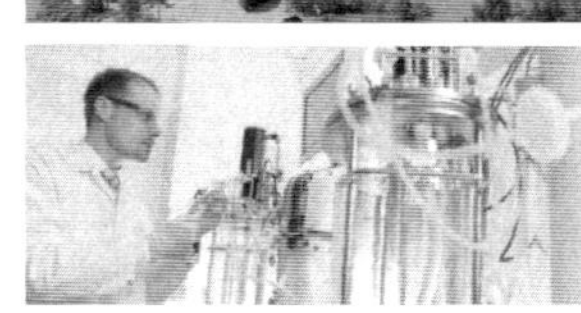

- Your independent partner since 1988
- State of the art technology (R&D and production)
- Fast, flexible, and tailor made technical service
  - **Sugazym HiTaA L** – high temperature stable amylase
  - **Sugazym MeTaA L** – medium temperature stable amylase
  - **Sugazym DX L** - dextranase
- Strong connection to sugar research institutes globally
  - E.g. Technical University of Berlin, poster from 2018:

    "New Approaches for the Determination of Dextran in the Sugar Production Process"

Technische Universität Berlin

Sugazym

Global sugar landscape

SternEnzym
The Enzyme Designer

Beet and cane sugar production

SternEnzym
The Enzyme Designer

Global beet and cane sugar production (1000 tonnes, raw value)

	2011/12	2012/13	2013/14	2014/15	2015/16	2016/17	2017/18
Beet sugar	39, 978	37,892	35,483	39,700	35,100	40,472	44,611
Cane sugar	137,028	145,322	146,369	145,000	134,900	137,526	147,539

Source: FO Licht (2018)

Global sugar production

SternEnzym
The Enzyme Designer

	World sugar production by region Oct/Sep (1000 tonnes, raw value)						
	2017/18f	2016/17	2015/16	2014/15	2013/14	2012/13	2011/12
EU	20,580	17,061	15,122	19,186	17,110	17,420	19,072
Europe*	33,867	30,344	25,913	30,343	27,656	28,598	30,837
Africa	12,522	11,351	11,519	11,404	11,301	10,964	10,295
N. & C. America	23,132	22,780	22,143	22,770	22,240	23,609	20,163
South America	48,160	48,083	46,529	42,826	47,389	48,685	43,016
Asia	70,122	60,679	61,499	68,841	68,471	67,072	66,131
Oceania	5,211	5,113	5,216	5,371	4,302	5,243	3,820
Total	**193,014**	**178,350**	**172,819**	**181,555**	**181,359**	**184,171**	**174,263**

***Includes EU + other countries in the continent including Turkey**
f = forecast

Source: FO Licht 2018 Note: May not add due to roundings

Trends in global consumption

SternEnzym
The Enzyme Designer

	World sugar consumption by regions Oct/Sep (1,000 tonnes, raw value)						
	2017/18	2016/17	2015/16	2014/15	2013/14	2012/13	2011/12
EU	18,732	18,702	19,346	19,254	19,267	19,090	19,018
Europe*	31,088	31,014	31,298	31,187	31,328	31,139	31,047
Africa	21,936	21,530	20,951	20,326	19,644	19,170	18,472
N. & C. America	21,599	21,256	21,021	20,791	20,663	20,335	19,541
South America	19,836	19,447	20,714	20,892	21,187	21,099	20,966
Asia	88,143	85,664	85,373	83,343	81,369	78,886	76,553
Oceania	1,630	1,636	1,730	1,714	1,706	1,701	1,696
Total	184,232	180.548	181,087	178,341	175,896	172,329	168,275

*Includes EU + other countries in the continent including Turkey
f = forecast

Source: FO Licht Note: May not add due to roundings

Per capita consumption

SternEnzym
The Enzyme Designer

Table 4. Per capita consumption of sugar 2017/18 (Oct-Sept, kg, raw value)		
	Per capita	Range
EU	36.35	22.75 (Portugal) -56.96 (Denmark)
Rest of Europe*	37.62	23.83 (Norway) – 45.07 (Russia)
Africa	17.60	5.54 (Ethiopia) – 47.36 (Sudan)
N & C America	37.06	34.00 (Canada) – 71.97 (Cuba)
South America	46.98	21.02 (Venezuela) – 56.79 (Brazil)
Asia	20.19	12.64 (China) – 73.31 (Israel)
Oceania	42.49	10.09 (Other Oceania) – 54.26 (Australia)
World	24.62	

*Includes EU + other countries in the continent including Turkey
Source: FO Licht 2018

Global supply/demand balance, million tonnes

SternEnzym
The Enzyme Designer

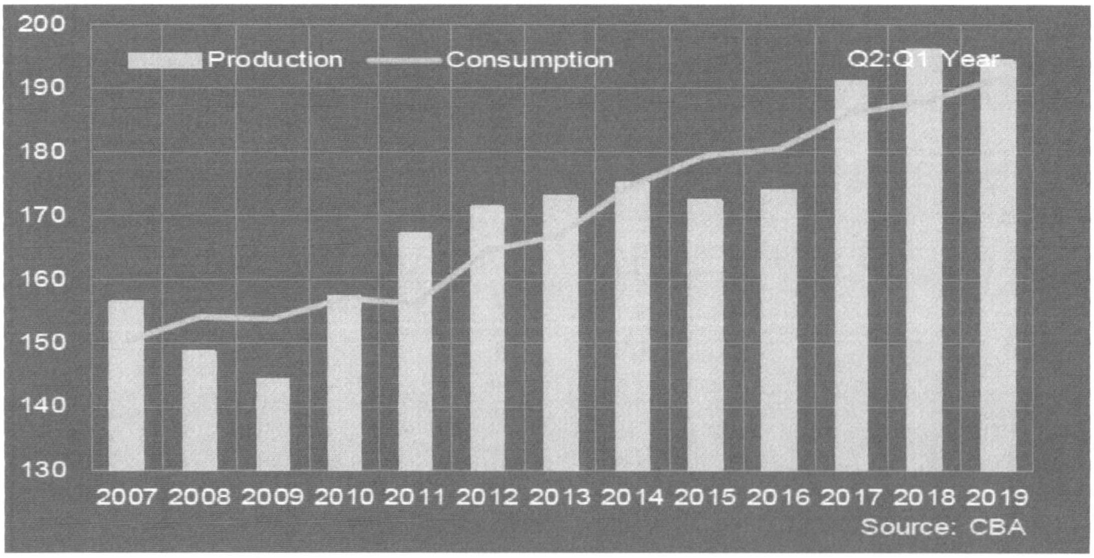

SternEnzym
The Enzyme Designer

Key drivers of supply and demand

- weather
- policy measures (to include national and regional that support trade barriers, quotas, subsidies, soft loans)
- speculation
- currency movement
- rising population, urbanization and changing taste,

and increasingly

- demonization of sugar by health campaign advocates

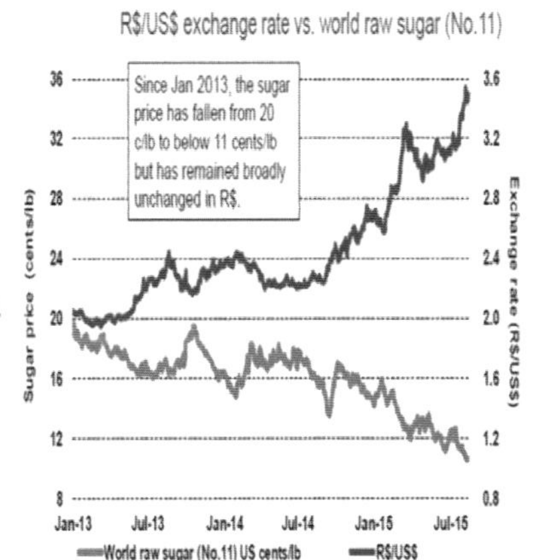

Source: LMC

SternEnzym
The Enzyme Designer

Policy and policy making

"I hear he turned water into wine, but it was a rather poor-quality Mesopotamian Cabernet."

Three children quarrelling over who should get the flute

SternEnzym
The Enzyme Designer

- **Anne claims the flute because she is the only one the three who can play it**
  - both Bob and Clara acknowledge this
- **Bob is so poor, he has no toys - therefore feels flute would give him something to play with**
  - Both Anne and Clara appreciate that Bob is not as well off as they are
- **Clara says she should have the flute because she has spent several months making it**
  - This is not disputed

Three children quarrelling over who should get the flute

SternEnzym
The Enzyme Designer

- **Anne claims the flute because she is the only one the three who can play it**
  - both Bob and Clara acknowledge this

Utilitarian

- **Bob is so poor, he has no toys - therefore feels flute would give him something to play with**
  - Both Anne and Clara appreciate that Bob is not as well off as they are

Economic egalitarian

- **Clara says she should have the flute because she has spent several months making it**
  - This is not disputed

No-nonsense libertarian

Elements sugar policy making

- efficient agricultural production that takes advantage of innovative technologies and practices
- adequate logistics infrastructure facilitating effective movement of cane/beet to factories
- Cane payment system that is fair to both growers and processors
- Effective collaborative structure in place whereby the growers, processors and researchers maintain dialogue to drive continuous improvements
- Domestic market protected from dumping
- Clear roadmap for the commodity driving future investments

Country case studies of policy making

Proactive – EU

- Reform of the sugar regime decreased output by some 5 mln tonnes
- Beet acreage decreased from 2,193,583 ha in 05/06 to 1,313,697 ha 15/16
- Inefficient producers incentivised to leave
- Currently 109 sugar plants operating, from 189
- Existing producers have become globally competitive
- 2017/18 output rose 23% to 20.5 mln t raw value
- EU exported 1 mln t in the first 3 months to Egypt, Mauritania, Sri Lanka, Israel, Turkey, Lebanon and Syria.

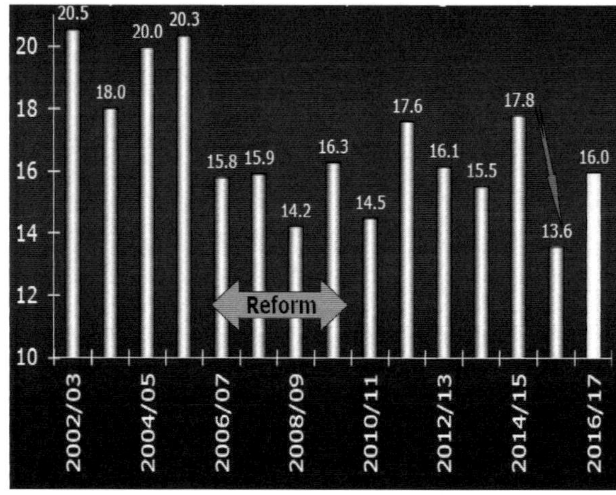

Sugar production (mln tonnes white value)

Source: FO Licht

Productivity gains driving competitiveness

Classification of cane producers by sugar yields		
Low yields lower than 5.42 t/ha	**Middle yields 5.42 to 9.38 t/ha**	**High yields higher than 9.38 t/ha**
Belize	Argentina	Australia
Bolivia	Brazil's NNE	Brazil's CS
Cuba	China	Colombia
Dominican Rep.	Egypt	El Salvador
Guyana	India	Ethiopia
Indonesia	Mexico	Guatemala
Jamaica	Pakistan	Honduras
Panama	South Africa	Malawi
Venezuela	Sudan	Peru
Vietnam	Thailand	Swaziland
	USA	Zambia
	Philippines	Zimbabwe

Source: International Survey of Agricultural and Industrial Yields of Sugar Crop MECAS(17)17. ISO, November 2017

Classification of beet producers by sugar yields (based on averages for 2013/14-2015/16)		
Low yields lower than 6.40 t/ha	**Middle yields 6.40 to 12.08 t/ha**	**High yields higher than 12.08 t/ha**
Belarus	Austria	Belgium
China	Canada	Chile
Egypt	Czech Rep.	Denmark
Finland	Greece	France
Moldova	Italy	Germany
Romania	Japan	Netherlands
Romania	Poland	Switzerland
Ukraine	Turkey	UK
	USA	

Source: International Survey of Agricultural and Industrial Yields of Sugar Crop MECAS(17)17. ISO, November 2017

Impact of EU sugar reform on the processing sector

SternEnzym
The Enzyme Designer

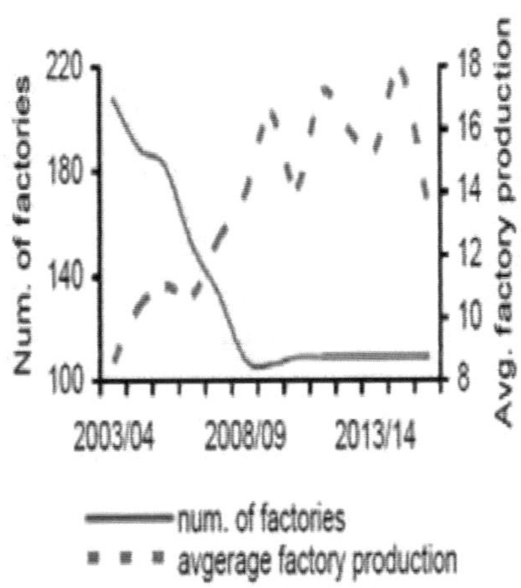

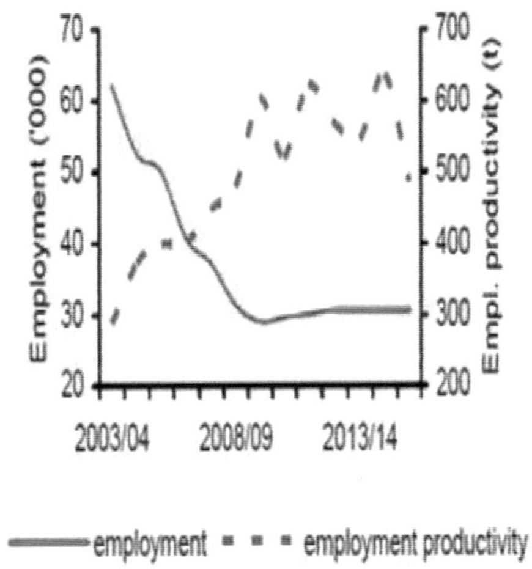

Reactive - Brazil

SternEnzym
The Enzyme Designer

- Rocketing oil prices in'70s prompted govt to act
- National Alcohol Programme ProlAlcool legislated in 1975
- Production of fuel ethanol from cane incentivised
- Between 1975/76 and 1990/91, cane output increased from 68.5 mln to 224 mln t
- During 1990-2010, sugar production increased 5-fold and exports 17-fold
- Credit crunch in 2008 has rationalised the industry through indebtedness
- Some 380 factories currently operating, down from 441

Sugar and ethanol production – 1990-2017

Year	Sugar production (mln tonnes)	Sugar exports (mln tonnes)	Ethanol production (bln litres)
1990	8.0	1.6	11.5
2000	16.5	6.5	10.6
2005	28.1	18.4	16.2
2010	39.0	28.0	27.2
2015	40.5	29.4	30.4
2017 (f)	37.1	25.1	26.2

Source: FO Licht

Estimated total debt in Brazil's sugar-ethanol sector

SternEnzym
The Enzyme Designer

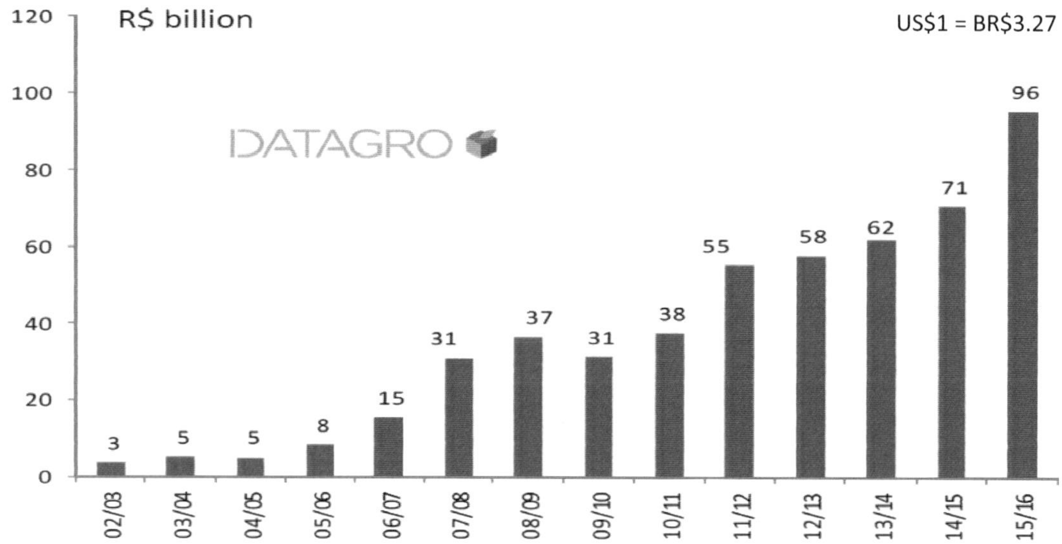

Political – India

SternEnzym
The Enzyme Designer

- 2nd largest sugar producer and leading consumer
- Some 50 million farmers grow cane on 5 mln ha
- Sugar is highly regulated of all commodities
- Cyclical sugar production cycle stemmed after removal of the market arm of policy
- Many politicians come from the sugar sector
- This gives way to "pork-barrel" politics
- Production policy measure continues to favour cane growers
- Sugarcane – most profitable compared to other crops

India sugar production and consumption (mln tonnes, raw value)

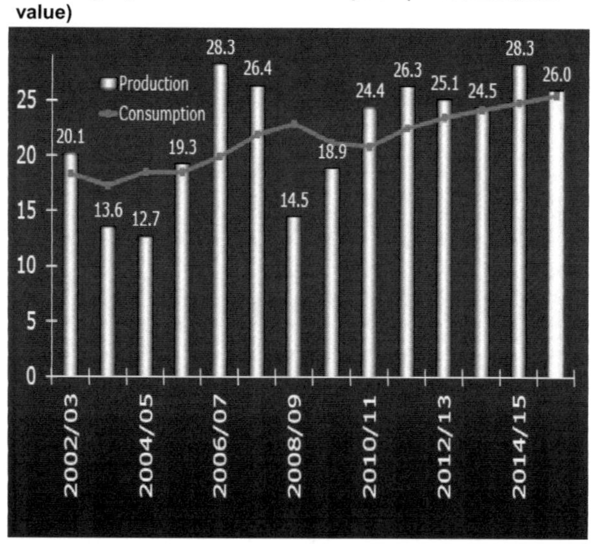

Source: FO Licht

Profitability of cane crop

SternEnzym
The Enzyme Designer

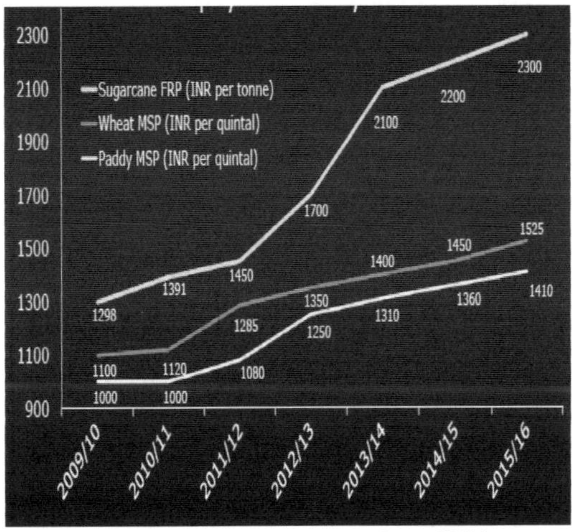

Source: FO Licht

Comparative analysis of profitability for cane growers in Uttar Pradesh, 2012/13

INR per hectare			
Crop	Expenses	Income	Net Income
Sugarcane	67,000	182,000	115,000
Rice/Paddy	30,000	58,000	29,6000
Wheat	26,600	52,500	25,900
Lentil/Pulses	24,250	63,000	38,750
Mustard	23,250	52,500	29,250
Potato	56,000	125,000	69,000

Source: Sugar mill input, Rabobank 2014

Apathetic – Caribbean Sugar Industry

SternEnzym
The Enzyme Designer

Sugar production the Caribbean (1000 t)

	17/18	16/17	15/16	14/15	13/14	12/13	11/12	10/11	09/10	08/09
Barbados	10	10	7	11	16	18	25	25	26	33
Belize	210	171	133	142	124	122	121	105	94	98
Jamaica	101	88	83	134	154	128	132	140	122	126
Guyana	140	140	183	231	216	187	218	237	221	234

Source: FO Licht

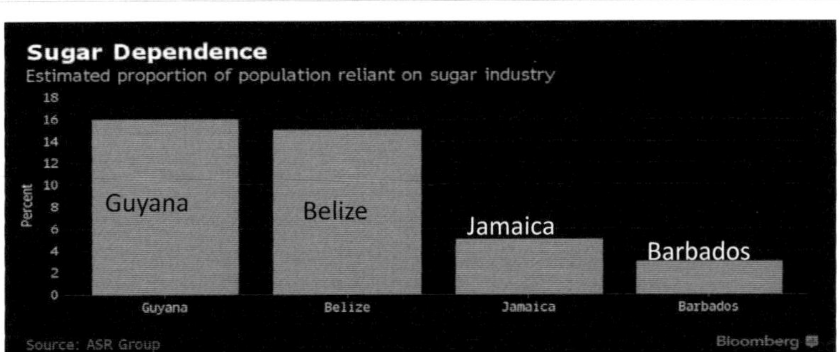

Concluding thoughts

SternEnzym
The Enzyme Designer

25 International Sugar Journal | Agribusiness intelligence

Concluding comments

SternEnzym
The Enzyme Designer

- Sugar industry not well served by thoughtful policy makers
- Political masters rarely in tune to serve the sector
- Effective policy making can make a difference, take EU

26 International Sugar Journal | Agribusiness intelligence

SternEnzym
The Enzyme Designer

27 International Sugar Journal | Agribusiness intelligence

International Sugar Journal
Agribusiness intelligence | informa

Informa
Christchurch Court
10-15 Newgate Street
London EC1A 7AZ

T +44 (0)20 7017 5000
E info@informa.com
W www.internationalsugarjournal.com
www.agribusinessintelligence.com

#1190 - Development of a Linear Pan Growth Approach (part ii)

Chris Mayhew, Carlton Haynes, Robert Howe, Craig Parker
British Sugar

The fine control of the mother liquor saturation is key to efficient crystal production. Over many years the industry has developed several methods to approximate the mother liquor conditions across the total cycle of the batch pan. Both total massecuite brix and mother liquor brix have been used individually to control the degree of supersaturation indirectly, but neither method can accurately guarantee control of crystal growth in isolation. It is the interaction between both these brix measurements that can be used to portray the bigger picture of how the crystal is growing.

This study, over the 2017/18 beet campaign at our Wissington Factory, builds on the details previously reported (paper #1153 – Linear Pan Growth, 2017 SIT conference, Taiwan) on the investigation into how combining the output from a microwave brix meter and a refractometer can be used effectively to improve crystal growth within a batch pan.

Introduction

Controlling pan growth conditions is a key component in delivering consistently good quality sugar from white batch pans. The ideal system for pan growth control is one which balances the rate of crystal growth with the saturation conditions within the batch pan. However, measuring these saturation conditions in a batch pan is not easy, as there are no instruments available that measure saturation directly. Therefore, pan control systems are based in part on measured parameters and part calculation. Knowledge of mass balances, sucrose solubility curves and pan measurement enables sugar technologists to derive a close approximation to saturation. This data can then be used to maintain the required growth conditions.

During the 2016-17 beet sugar campaign, British Sugar embarked on a project, at its Wissington factory, to improve the sugar quality in its white batch pans. The system model, developed for British Sugar by Chris Rhoten, was based on the principles of Linear Growth[1]. The Linear Pan Growth (LPG) model involves the measurement of massecuite and mother liquor brix which, when combined with the pan temperature, is used to control the addition of the feed syrup to control the **D**egree of of **SAT**uration (D_{SAT}) profile. This in turn creates a massecuite brix profile and mother liquor brix profile for the pan cycle.

The project demonstrated syrup purity and crystal quality improvements when the white batch pans used the LPG control profiles. However the gains were not delivered consistently when deployed in three of the six white batch pans at Wissington. Closer examination of the pan instrument profiles highlighted inconsistencies in mother liquor brix profiles. This appeared to relate to changes in the crystal content and particle size of pan seed supplied by the Cooling Crystalliser system. This finding initiated an additional work stream, aimed at understanding and improving the performance of the Cooling Crystalliser seed system. The outcome of that piece of work has been reported in a separate paper at the 2018 SIT conference[2].

During the first quarter of the 2017-18 beet sugar campaign, the final modifications to the Cooling Crystalliser seed system were commissioned, trialled and proven to deliver a more consistent seed quality for the white batch pans. The focus of the team then reverted back to developing the LPG project.

Pan profiles

The first element of work to be revisited in 2017-18 was the development of an open brix profile that is a key part of the LPG profile approach.

- ***Brix profile control***

Initial work in 2106-17 focussed on developing a Delta brix strategy. The aim was to control the difference between the massecuite brix and the mother liquor brix in the white pan. However, the team struggled to achieve repeatability between pan cycles. This issue was associated with a slow drift on the microwave brix meters and a regular three week calibration program was required to keep their data valid. At the same time, calibration checks on the pan refractometers confirmed that they remained within their $\pm$ 0.2 brix performance window. The challenge confronting the team was the reliance on measuring and controlling the delta brix profile to match the LPG model's requirements in the very early stages of the pan cycle post seeding. Ultimately the extremely small difference in delta brix at this point was beyond the combined accuracy of the two measurement techniques. Consequently, the team were unable to deliver consistent profiles (this was alluded to in the previous LPG paper presented in 2017).

The team therefore explored alternative routes to obtain the measurement information required to support the LPG approach. The first approach taken was to reconfigure some batch pans to have microwave brix or refractometer brix as the primary control, as defined from the LPG model.

- ***Steam profiles***

The original Linear Model, developed by Chris Rhoten, defined a mother liquor saturation target of 1.2 D_{SAT}. This saturation then defined the linear crystal growth rate. The model assumed a fixed calandria steam pressure and a steady vacuum within the batch pan, which defined the rate of evaporation. The model controlled the target D_{SAT} by the addition rate of the syrup feed, with the mother liquor brix and massecuite brix profiles generated as the outcome of the model.

However, the steam profiles for the batch white pans at Wissington are not fixed. They are set to ramp up, accompanied by a slow ramping down of the pan vacuum. This was originally implemented to maintain an even temperature profile over the life of the pan, with a view to:

- Minimise colour formation that would occur if the massecuite temperature rose
- Allow a lower grade vapour to be used over the pan cycle to save energy.

When the revised profiles were put into the model, it started to highlight a key factor affecting the saturation to growth rate balance.

- ***Evaporation vs growth rate***

An underpinning principle of LPG control relates to the mother liquor saturation vs the linear growth rate. In the author's previous paper - Development of a Linear Pan Growth Approach., Mayhew, Haynes et al[3], it is reported that the linear growth models, developed by Chernov A.A.[4], are dependent on the D_{SAT} of the mother liquor. The relationship is linear over the purity range experienced in white pans.

Under typical operational conditions, pan steam pressures vary leading to changes in evaporation rates. These changes lead to variations in the mother liquor D_{SAT}. If the steam pressure is low it reduces the crystal growth rate and the pan slows down. However, if the steam pressure increases it drives up the D_{SAT} which could lead to false grain forming, if the crystal growth rate is unable to match the evaporation rate.

At all of British Sugar's factories, the white pan speeds are set through the target steam pressure they aim to achieve during the feed phase of the pan. They ramp from the post seeding hold pressure (typically 900mbara), to a final pressures in the range from 1000 to 1180mbara (during campaign operations). The original LPG model had a steam pressure target of 1200 mbara. However, with the actual final steam pressure achieved in the pans falling below the model's target, it became apparent that the evaporation rate from these pans was not sufficient to maintain the mother liquor's D_{SAT} profile to match the modelled growth rate.

This observation indicated that the traditional practise of operating at different steam pressures to achieve production targets, would require a different D_{SAT} profile to balance with the relevant rates of evaporation. Therefore, if the steam pressure profile is altered to deliver throughput requirements, then the pan profile should be amended to balance the growth rate to the achievable mother liquor D_{SAT}.

- ***Development of the online D_{SAT} calculation***

During the 2017/18 campaign, an online D_{SAT} calculation was developed using the brix data from the pan refractometer and microwave brix meter, combined with the pan temperature measurement. The calculation uses the brix meter data, combined with the feed syrup purity, to calculate the crystal content within the massecuite, using a solids balance. A sugar balance then enables the mother liquor purity to be calculated, which when combined with the massecuite temperature, is used to calculate the D_{SAT} of the mother liquor. The calculated D_{SAT} was added as a parameter within the Distributed Control System (DCS) and the pan graphical trends.

The use of the calculated D_{SAT} to influence pan profiles

Once the D_{SAT} parameter was added to the individual trends for pans 4, 5 and 6, the team could monitor the mother liquor D_{SAT} through the full pan cycle, an example of which can be seen in Figure 1. Once the mother liquor D_{SAT} trends were available it highlighted a range of problems affecting the pan performance:

- the brix and temperature conditions from when the pan was charged through to the approach to seed point
- changes in the mother liquor conditions immediately post seeding
- during the feed and brix up stages.

- ***Adjusting the approach to the seed point***

Under the original pan profile, the pre-seed boil down phase was quite aggressive, cutting the steam pressure back from ~1200 mbara to 800 mbara, 0.3 brix before the seed point target (80.5 brix). From the D_{SAT} information it could be seen that with this approach, the target mother liquor D_{SAT} at seeding (1.15) was rapidly exceeded, often rising to 1.17 prior to the start of the feed phase. With visibility of the D_{SAT} information, the seeding point was reduced to 80.3 brix, to target a 1.13 seed D_{SAT}. This has been proposed as an appropriate target by

Rozsa et al[5]. At the same time, the post-seed steam pressure profile was reduced to prevent the mother liquor D_{SAT} rising above the target seed point D_{SAT} during the seed establishment phase. Both these changes reduced the visual observation of false grain appearing on the pan cameras during the seeding phase.

- ***Changes to the steam profiles during the feed phase***

Further changes were made to the steam pressure profile post seeding, to maintain a flat mother liquor D_{SAT} profile during the first third of the feed phase. The initial feed phase steam pressure was reduced from 1000 mbara to 900 mbara and the steam ramp time was extended by 15 minutes. The changes were made to control the evaporation rate during the early stages of the feed phase. This was to maintain a mother liquor D_{SAT} profile that matched the growth rate to the available surface area of the seed crystals. The modification was successful in suppressing the D_{SAT} rise during the first third of the feed phase, as seen between the profiles in Figures 1 and 2. Visual observations, made through the pan cameras, confirmed a significant improvement in the number of the crystal present post seeding, with no visual evidence of false grain forming during this stage of the pan profile.

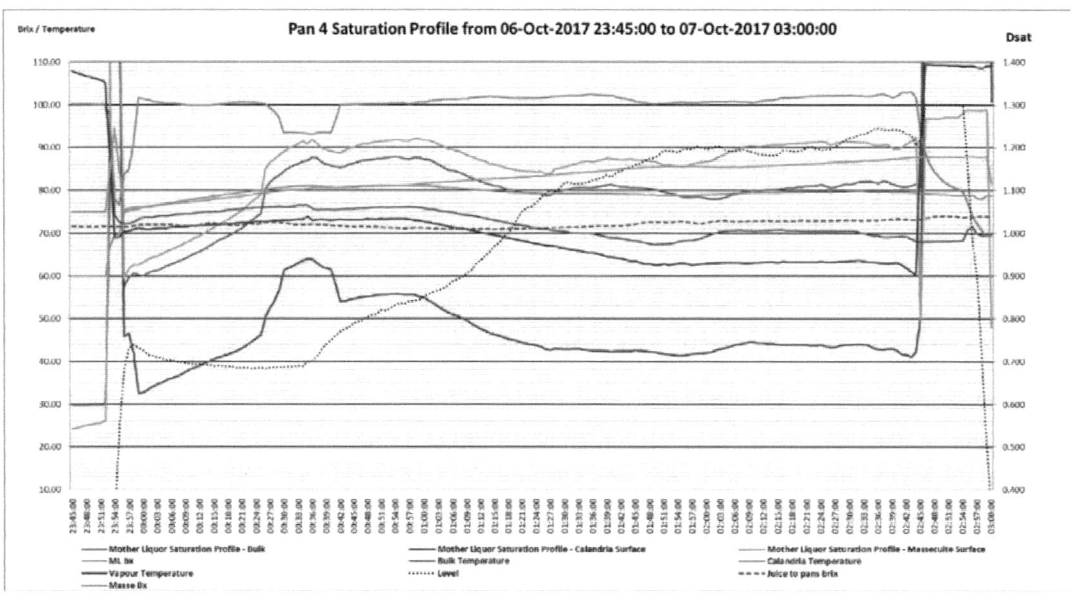

Figure 1 - Original pan 4 profile with D_{SAT} calculation active

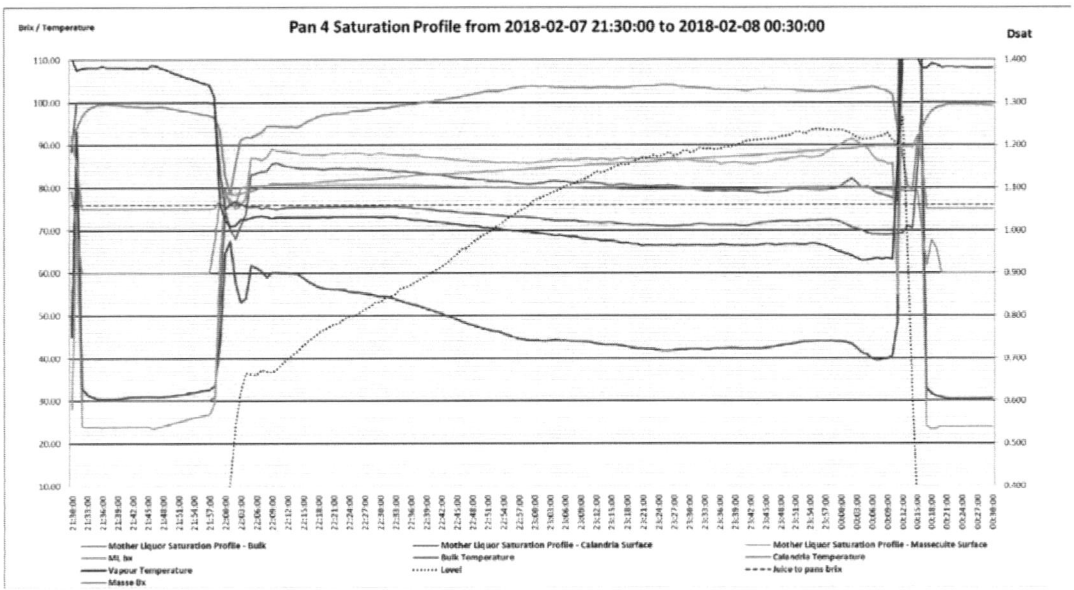

Figure 2 - Revised pan 4 profile with revised start point and longer steam ramp

- ***Linking trends and observations***

One of the key advantages that the team realised, once the D_{SAT} profiles were added to the pan parameter trends, was the rise and fall of the D_{SAT} profile during the pan cycle. The challenge recognised by the team was to link the information with key events and observations from the other pan instruments. This then had to be related to possible influences that could impact these trends. A typical example, shown in Figure 3, shows a rise in the bulk mother liquor D_{SAT} post seeding. After approximately 35 minutes into the feed phase the bulk mother liquor D_{SAT} reaches 1.15, before falling back down. Just after the D_{SAT} peak was recorded, fine grain could be visually seen appearing on the pan camera. This repeated itself for the next two cycles, before a change in the feed control profile was instigated that flattened the D_{SAT} profile post seeding. This eliminated the false grain appearing on the pan camera at the same point in the pan cycle.

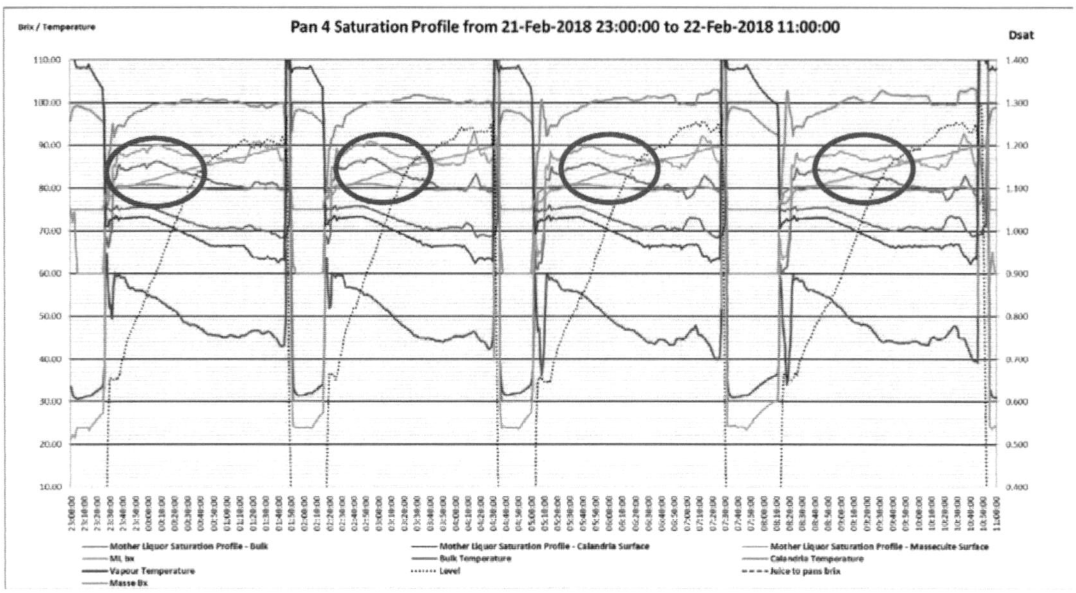

Figure 3 - Modified pan 4 profile to control the D_{SAT}

In the above example, the issue was the imbalance between the surface area of the seed crystal and the targeted rate of growth of the pan profile. This highlighted the problems experienced by the Wissington team when operating their white pans to an overall massecuite brix profile (i.e. just using microwave data). The overall massecuite brix measurement cannot discern between dissolved solids and crystallised solids. In the case highlighted from Figure 3, the seed crystal surface area was insufficient to remove the sucrose from solution at a suitable rate to match the evaporation rate determined by the target brix profile. This resulted in the mother liquor brix rising, forcing the D_{SAT} higher, until the system rebalanced itself with a crop of false grain as it reached a D_{SAT} of 1.15.

The fourth pan profile seen in Figure 3, shows the effect of a change in the target massecuite brix profile, which had been made to overcome the rising D_{SAT} in the first 35 minutes of the feed phase. The action was to slightly lower the massecuite brix profile over this period in the pan, thus rebalancing the evaporation rate with the crystal growth rate. The impact can be seen in the flattening of the D_{SAT} profile driven by the mother liquor brix being around 0.3 brix lower at this point in the cycle.

Data collection

At Wissington the 6 white batch pans are arranged into two streams of 3 pans, with A-side being pans 1, 2 & 3 and B-side consisting of pans 4, 5 & 6. Each stream runs separately through a dedicated receiver, continuous vacuum crystalliser (CVC), associated batch centrifuges and granulator before merging together prior to the sugar cooler. All 6 white pans operate from a common feed syrup system and also share a common steam supply and vacuum system.

Following the results obtained from the 2016/17 LPG study, the team undertook to improve the performance and reliability of the Cooling Crystalliser system at Wissington. The improvements were made during the first half of the 2017/18 campaign and therefore had an influence on the performance results from both sets of batch pans involved with this part of the LPG study. It was therefore decided to overlay those changes onto the data collected for the LPG project, to enable a representative comparison of the LPG controlled pans to be made against the standard British Sugar pan profile.

Based on the evaluation made in the 2017, sugar samples were taken from A side and B side granulators and analysed for MA (Mean Aperture) and CV (Coefficient of Variance). These samples were taken on a daily basis and were cross-referenced against the changes made to the Cooling Crystalliser system. Syrup samples, from No.1 BMA centrifuge (fed from A side pans) and from No.4 BMA centrifuge (fed from B side LPG controlled pans), would be taken during a period when the improvement work on the Cooling Crystalliser system was complete.

Results

The activities to develop the performance of the Cooling Crystalliser, improved the repeatability of the particle size regularity and crystal content of the seed used in the white batch pans (Figure 4). The big improvements can be seen associated with the stirrer modification and relocation of the Ditmar seed insertion point. The variation of the final seed MA in the last quarter of the chart in Figure 4, shows the output of changing the charge liquor brix and its influence on the seed MA produced.

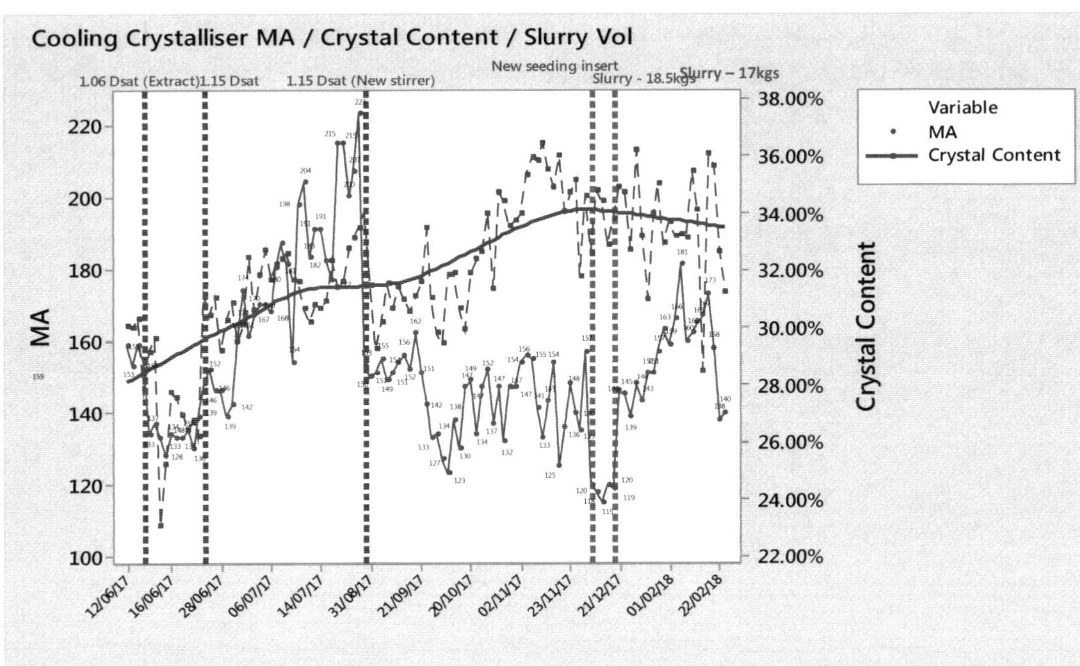

Figure 4 - Cooling Crystalliser Product Quality

Both A and B stream pans benefitted from the increasing crystal content and improved MA regularity of the seed slurry from the Cooling Crystalliser work. The evidence of this improvement is shown in a reduction in the CV of the sugar samples from both A and B side granulators over the 2017/18 campaign, as shown in Figure 5.

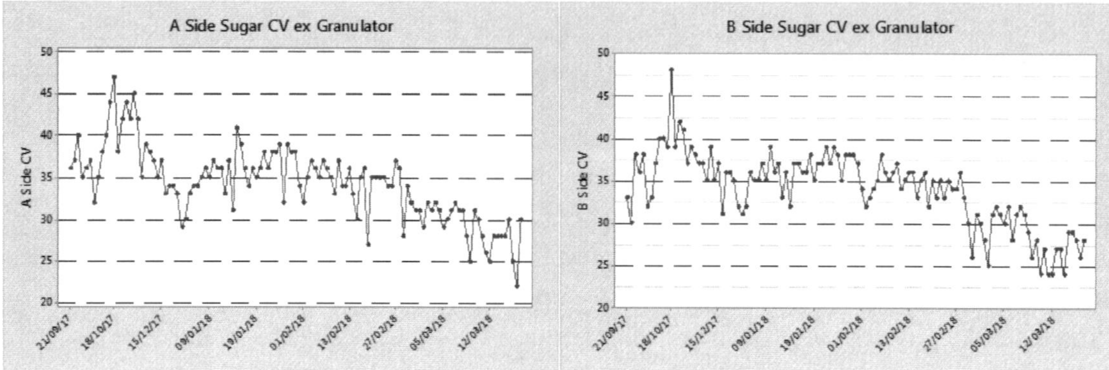

Figure 5 - Impact of seed system improvements on Sugar quality from both A side and B side Batch Pans

The impact of the seed volume changes to the white pans became evident as soon as the saturation calculations were added to the DCS pan trends. Figure 6 shows a trend from pan 4, during the first half of the campaign, where work was focussed on improving the crystal quality of the Cooling Crystalliser system. The trend show the pan seeding correctly at 1.14 D_{SAT}, operating at a reduced steam pressure to allow the grain to establish and then slowly ramping up the steam pressure as per the revised control program. In the trend, the D_{SAT} can be seen to fall for the lower steam pressure period after seeding (point A), before rising to point B as the steam pressure increases. At this point, the evaporation rate is exceeding the crystal growth rate. The system corrects this imbalance by creating additional seed sites (false grain) to increase the overall crystal surface area. This occurs such that the crystal growth rate will match the evaporation rate. The D_{SAT} can then be seen to fall away as the system now has more surface area than to match the chosen steam profile.

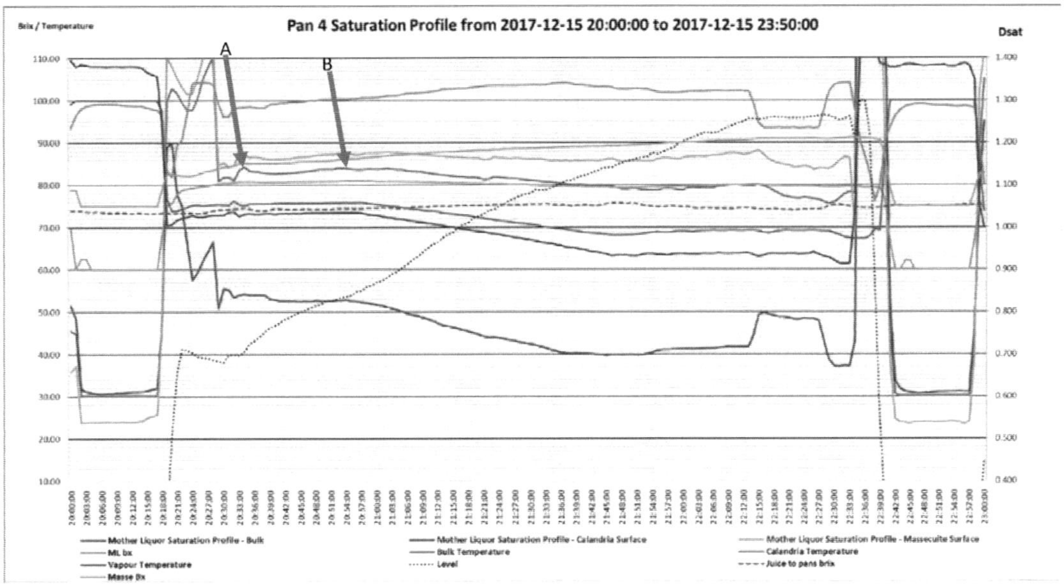

Figure 6 - Pan 4 Saturation profile (Insufficient seed volume added)

Figure 7 shows the influence of an increased seed volume on the D_{SAT} profile. Here, sufficient surface area is added during the seeding step and the D_{SAT} is seen to hold a shallow decline over the full feed phase of the pan cycle. In this situation the growth rate, as driven by the D_{SAT}, is in close balance with the steam pressure profile of the pan (hence the evaporation rate), allowing the crystal to grow quickly but within the control of the pan program.

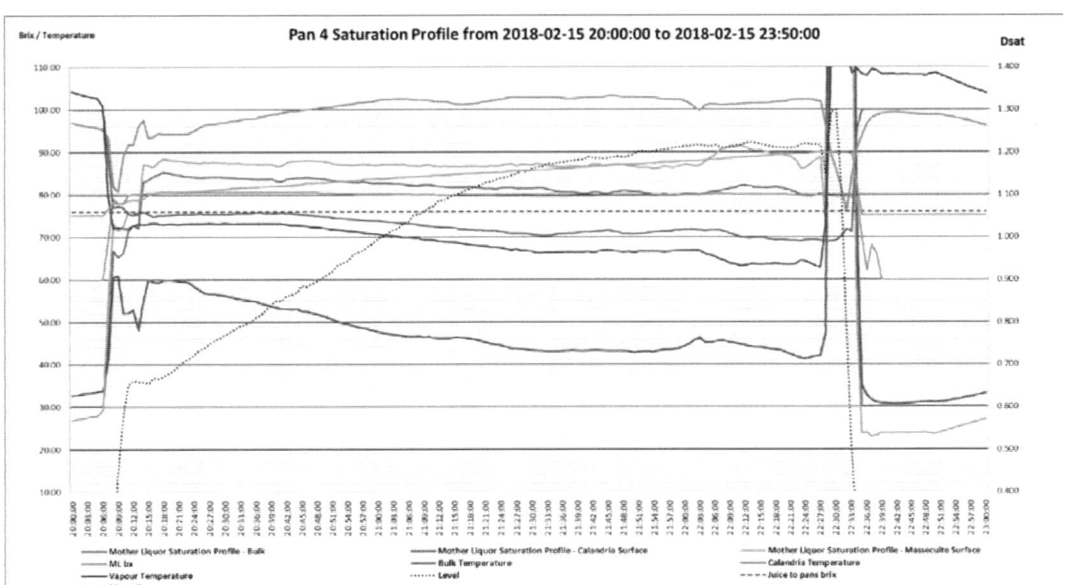

Figure 7 - Pan 4 Saturation profile with increased seed volume

- ***Replicating B side steam profiles onto A Side pans***

The impact of the rapid rise in steam pressure on the mother liquor D_{SAT}, as shown in Figure 1, raised concerns about how this may be affecting the mother liquor D_{SAT} on the A side pans (i.e. not running under the LPG profile). These pans were not fitted with pan refractometers, however it was possible to observe a rise in the massecuite temperature in pan 1 (Figure 8). With a consistent vacuum profile, a rise in the pan temperature can be related to a rise in the mother liquor brix. Comparing this temperature profile rise with the flatter profile from Pan 4 (Figure 7) would suggest that the D_{SAT} in pan 1 was increasing post seeding and then falling back. This hypothesis was further supported by reviewing the early trends observed on Pan 4 (Figure 1), which exhibited the same temperature and D_{SAT} relationship.

Following this observation, the post seeding steam profiles for A stream (pans 1, 2 and 3) were changed to match the ramps created for B Stream (pans 4, 5 and 6). The impact of this change can be seen in the revised pan 1 profile (Figure 9), with a flatter temperature profile post seeding matching that seen in pan 4.

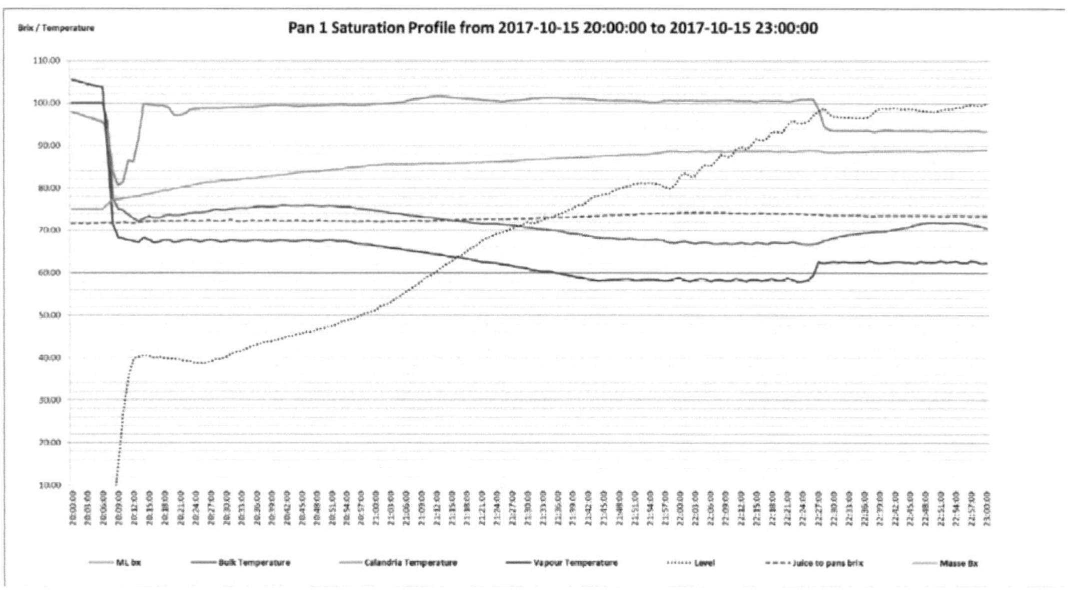

Figure 8 - Pan 1 Original Steam Profile

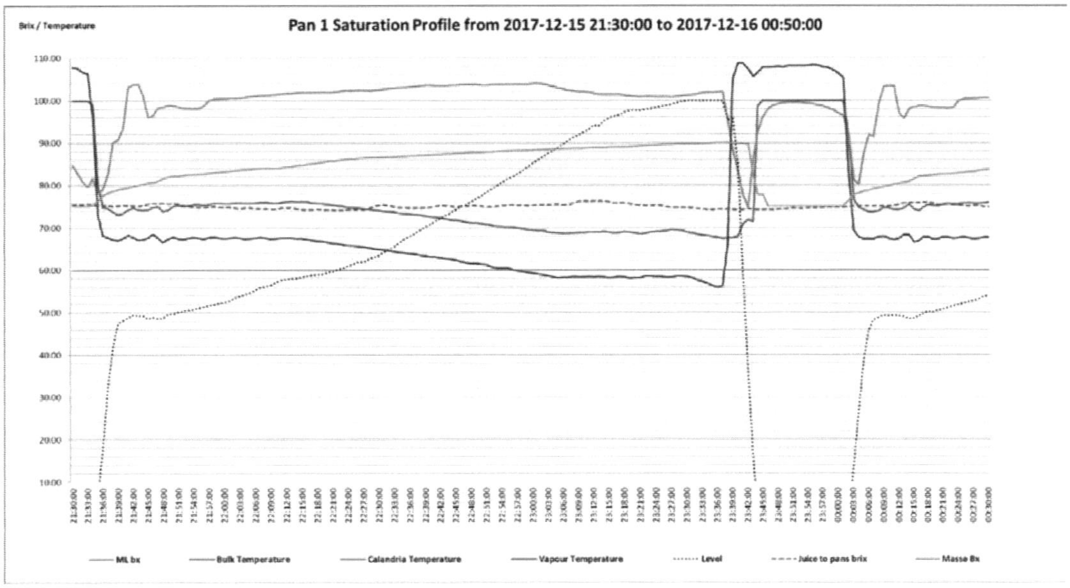

Figure 95 - Pan 1 with revised steam profile

- ***Comparison data A side vs B side for MA, CV and Syrup Purities***

Once the improvements to the Cooling Crystalliser system had been made, consolidated and understood, the evaluation process focussed back onto the LPG program changes on the white batch pans. A short trial period, running for two weeks in February 2018, focussed on collecting sugar samples from both granulators as well as syrup samples from No.1 BMA centrifuge (A side pans) and BMA No.4 (B side pans). The results of this analysis, shown in Table 1, compare the trial period in the 2017/18 campaign, with the data collected in the 2016/17 campaign.

	Stream	Pre-16/17 Campaign trial	16/17 Linear Growth Trial	17/18 Linear Growth Trial
Syrup Streams				
High Green Purity	A Side	84.17	83.43	87.00
	B Side	85.15	83.48	86.90
	Difference	***0.98***	***0.05***	***-0.10***
Wash Syrup Purity	A Side	87.39	86.14	91.40
	B Side	87.89	85.65	90.80
	Difference	***0.5***	***-0.49***	***-0.60***
Feed Syrup	***To pans***	94.54	94.02	95.63
MA of Sugar Produced (μm)				
Ex Granulators	A Side	564.0	562.3	609.7
	B Side	549.9	547.1	584.9
	Difference	***-14.1***	***-15.2***	***-24.8***
CV of Sugar Produced				
Ex Granulators	A Side	34.2	34.4	31.7
	B Side	34.9	34.1	31.3
	Difference	***0.7***	***-0.3***	***-0.40***
% Through 224 (μm)				
Ex Granulators	A Side	2.9	3.1	1.6
	B Side	4.0	3.3	1.7
	Difference	***1.1***	***0.2***	***0.1***

Table 1 - Analysis comparison for LPG vs. Standard profiles across the 2016/17 and 2017/18 trial periods

From the results in table 1, there was a small improvement in the average purity in both High green and Wash syrup between A-side and B-side. It must be noted that the difference in the average high green purities is within the analytical repeatability of the laboratory test method and therefore this result cannot be held as statistically significant. However, it still confirms the improvement of approximately 1 purity point when compared with the pre-trial base case.

The wash syrup purity difference is at the limit of the laboratory test method's analytical repeatability, with a similar small improvement on the difference also seen between A-side and B-side. Again the analysis confirmed a 1.1 delta purity improvement between the 2017/18 trial and the 2016/17 pre-trial base case.

The Ex Granulators sugar CV's also continue to exhibit the improved difference between the A-side and B-side streams, with an average 0.4 CV difference (in favour of the LPG controlled B-side pans). This confirms a 1.1 points overall improvement in the sugar CV when compared with the pre-trial base case. The table also shows a reduction in the overall average CV from 34.3 to around 31.5 during the 2017/18 trial period, due in part to the improvements made to the Cooling Crystalliser system earlier in the 2017/18 campaign.

Ongoing improvements

- ***D_{SAT} control***

The next step for the LPG program development is to create a D_{SAT} controller for the pans, using the syrup feed rate to control the calculated D_{SAT} result to follow the target D_{SAT} profile. This will increase the reliance on the output of the four measurement devices;

  - Microwave brix meter (massecuite brix)
  - Refractometer (mother liquor brix)
  - Pan temperature probes
  - Pan level transmitter

The first three devices contribute directly to the D_{SAT} calculation, whilst the fourth (level) give us the reference point to control the D_{SAT} profile to.

In terms of reliability, the temperature probes are the most basic, but robust, part of the measurement components. There are three probes, all spaced evenly around the bottom cone of the pan, on the same level as the microwave meter and refractometer. These temperature probes are constantly cross-compared against each other, by the pan program, and any errors are identified and rectified automatically.

The reliability of the refractometers has continued to remain within the manufacturer's tolerance of $\pm$0.2 brix, with their on-board monitoring continuing to report a healthy status.

The microwave meters do exhibit some drift over the operating period, but a program of cross checking and re-calibration (if required) every 3 weeks has been recommended and adopted.

The output from the level transmitters is regularly compared with a visual check on the pans and occasional minor re-calibrations are required to re-range these devices.

The impact of the variance in either one of the brix meters will have a direct effect on the calculated D_{SAT}, and hence pan quality control. For example; an increase in the difference between the two brix meters will cause the calculated D_{SAT} to reduce and the feed control loop to cut back on syrup addition to the pan. This would result in the true D_{SAT} rising and the potential for false grain to form. A simple cross check, to highlight this issue, would be to observe the temperature profile of the pan. This would rise above its normal profile for the pan cycle, indicating a rise in the mother liquor brix. If this rise is also shown by the refractometer, it can be deduced that the microwave meter is drifting and requires re-calibration. Alternatively, if the refractometer profile remains flat, or drops away from the normal trend, it would indicate that the refractometer was drifting and should be cleaned and recalibrated.

A failure in the temperature monitoring will also impact on the D_{SAT} calculation, as it relates to the solubility element of the D_{SAT} calculation. However, a failure, either partial or total, on one of the temperature probes will be captured by the pan program and the output of this probe will be automatically excluded from the calculation. A fault will be registered, but this fault would not immediately impair the D_{SAT} calculation in this situation.

A drift in the level measurement will have a lesser impact on the D_{SAT} control, essentially as the D_{SAT}, steam and vacuum profiles are all related to the level of the pan and all will adjust in proportion to each other. The main issues with a level drift would be related to the capacity of the pan if the level drifts lower or potential sugar loss to the condenser system if the level drifts higher.

Conclusions

Operating batch pans using the LPG approach has continued to deliver the improvements in sugar CV from the granulators, high green purity and wash syrup purity that were first measured during the 2016/17 campaign. The work on improving the quality and consistency of the seed system has significantly improved the repeatability of both A-side and B-side batch pans, resulting in an overall reduction in the sugar CV from the base case 34.3 average down to 31.5.

- Providing a stable and consistent seed massecuite is a key ingredient in the development of a successful LPG approach

The open LPG feed profiles running during the early feed phase of the batch cycle, developed from the LPG principles, have been reinforced with the development of the mother liquor D_{SAT} calculation. The resultant D_{SAT} profiling identified issues with mother liquor variability that may have contributed to periods of false grain formation. These issues were subsequently addressed with changes to the brix profile and also modifications to the steam pressure profile over the pan cycle. The result of these changes produced a more even D_{SAT} profile over the feed phase of the pan cycle, driving smoother crystal growth and reducing the potential formation of false grain.

- Calculating the mother liquor D_{SAT} in real time, using total massecuite brix, mother liquor brix and massecuite temperature, unlocks valuable information on the massecuite conditions. This opens a window onto improved crystal quality production through better massecuite control during the growth phase.

Creating a feed control brix profile, to manage the mother liquor D_{SAT} in the absence of a DSAT control parameter, was a significant step in preventing false grain forming during the pan cycle. However, the system must also be balanced between the evaporation rate, which drives the DSAT profile, and the crystal growth rate, which is limited by crystal surface area during the early feed phase.

- The batch pan speed is intrinsically linked to the D_{SAT} profile, going faster means sailing closer to the D_{SAT} limit that prevents false grain production.

The key to an accurate D_{SAT} control profile relies on the accuracy of the field instruments. In this case these are: refractometer brix meter, microwave brix meter and temperature probes. Keeping the output of these instruments accurate is a major factor in the reliability of the

D_{SAT} calculation. Therefore regular calibration checks are essential to keeping the D_{SAT} valid and to manage the optimum crystal growth rate.

- Reliable massecuite brix and mother liquor brix meters are a key step towards measuring and controlling the mother liquor DSAT within the pan.
- Regular brix meter calibration checks will allow the process control required to deliver improved crystal quality control.

Acknowledgements

The authors would like to thank the Wissington factory operational team for their support and resources used during the trial

References

1. Linear Growth Rate Based Adaptive Batch Crystallisation Control, Christopher D Rhoten, ASSBT Proceedings from the 38th Biennial Meeting, February 23 - February 27, 2015

2. Developments in the repeatability of Cooling Crystalliser performance, Chris Mayhew, Carlton Haynes, Robert Howe, David Cooper and Craig Parker, British Sugar. SIT Proceedings from the 77th Technical Conference (Florida) 2018

3. Development of a Linear Pan Growth Approach, Chris Mayhew, Carlton Haynes, Robert Howe, Craig Parker and Chris Rhoten, British Sugar and Escon America Inc. SIT Proceedings from the 76th Technical Conference (Taiwan) 2017

4. Chernov A.A. The spiral growth of crystals, Soviet Phys - USPEKHI 4, 116-148 (1961)

5. Rozsa, L., Rozsa, J., Kilpinen, S., 2017, 'Crystal growth and crystallization control tactics in industrial sugar crystallizers, Part 2: crystal growth', International Sugar Journal, pp260-266

#1191 - Improving the repeatability of low-side sugar colour management

David Cooper, Mark Lomax, Stephen Owsten, Craig Parker and Bjarne Christian Nielsen
British Sugar and Neltec Denmark A/S

Traditionally, British Sugar has run its low side centrifuges at fixed massecuite load levels, with ratio control of wash water addition to achieve the desired colour of sugar produced. However, visual interpretation by the operators of sugar colour varies widely, leading to different water addition ratios being employed. This leads to variability in the colour of the sugar produced from the centrifuges, with its impact on colour recycle, as well as excessive water dilution at other times.

This paper reports on British Sugar's experience during the 2017/18 campaign at its Bury St Edmunds factory, using Neltec's ColourQ 1700 CC colorimeter system. The ColourQ 1700 CC system provided insight into sugar colour patterns across the centrifuge screen and how they changed due to a number of influences. The system was given full control of water addition and the trial data demonstrates its effectiveness in controlling the sugar colour to a much tighter standard deviation than the traditional approach employed by British Sugar.

Introduction

The fine control of solution colour from sugar discharged from continuous low side centrifuges is a key factor in ensuring refinery extraction is maximised. This is a fine balance between minimising sugar loss into the run-off syrup and minimising the recycle of impurities with the crystal.

Traditionally in British Sugar's factories, the feed rate of massecuite into continuous low side centrifuges would be set by fixing the machine's motor load to a desired target. The wash water flow rate would then be manually adjusted to achieve the desired sugar solution colour. However the operators tend to over wash the sugar to operate on the side of caution, thus increasing the sugar passing forward with the run off syrup and increasing refinery costs.

British Sugar's current operating strategy for continuous centrifuges is to utilise Distributed Control System (DCS) capabilities and ratio wash water addition to the load (massecuite feeding rate) of the machine. This allows the sugar solution colour to be automatically controlled as the centrifuge load alters to meet normal process fluctuations.

British Sugar wished to improve the variation in solution colour produced from its continuous centrifuges. To investigate what improvements could be realised a trial was conducted in partnership with Neltec Denmark A/S, using one of Neltec's ColourQ 1700 CC colorimeters on a raw sugar (2nd boiling) continuous centrifuge.

Continuous centrifuge colour control

As discussed above the control of solution colour of the sugar discharging from a continuous centrifuge has developed over the years from a fixed wash water approach to one of controlling the ratio of wash water flow to the load of the centrifuge. This has improved the quality of the final sugar discharged from the centrifuge, as well as reducing the sugar lost into the syrup run-off.

Setting up the target sugar solution colour being discharged from the continuous centrifuge is achieved by taking regular samples to the laboratory for analysis and adjusting the wash water flow until the target solution colour is achieved. However, in practice is not the full story as operators use their experience to tune the discharging sugar's solution colour by eye.

As the visual perception of sugar solution colour varies from individual to individual, the authors decided to investigate what level of variation was being introduced into the refinery operation. The simple experiment involved taking samples of the discharged sugar from a continuous centrifuge and determining the solution colour via laboratory analysis. Five operators were then asked to estimate the solution colour of the sugar samples using their experience (i.e. their eyes). Figure 1 below shows the level of variation in "perceived" solution colour for a number of sugar samples in the solution colour range of 875IU to 1530IU. As can be seen this variance could be as large as 700IU for a single sample.

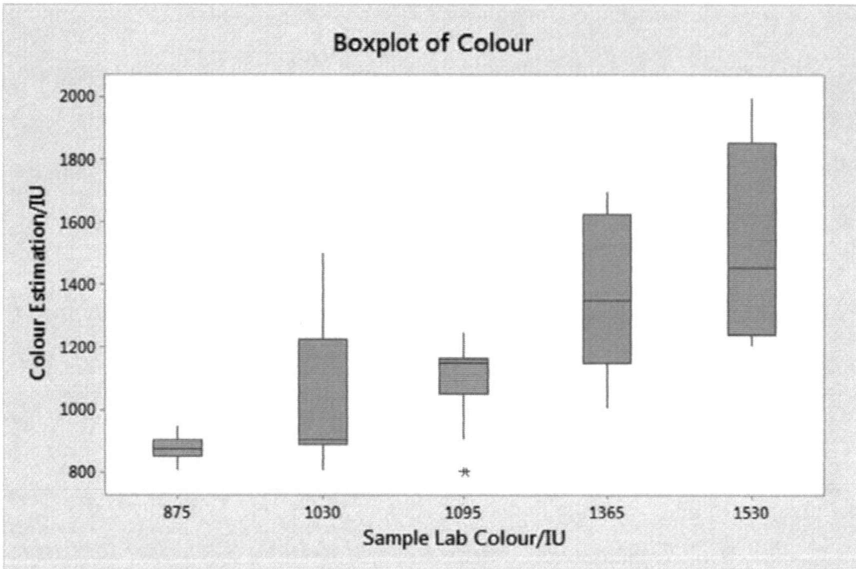

Figure 1 – Variance in operator's visual estimation of solution colour

ColourQ 1700 CC installation

A ColourQ 1700 CC colorimeter was mounted onto one of Bury St Edmunds' raw sugar continuous centrifuge. The installation required the removal of an access segment, used to change the centrifuges screens. This segment was replaced by a segment with the instrument attached (see Figure 2).

Figure 2 – ColourQ 1700 CC mounted on the continuous centrifuge lid

The mounting of the ColourQ colorimeter on the continuous centrifuge lid, provides an unobscured view of the internal basket area as it rotates. This location enables the instrument to monitor the sugar separation from the mother liquor, as the massecuite travels up the basket screen area. Figure 3 shows a schematic of the measurement beam and its orientation to the basket area.

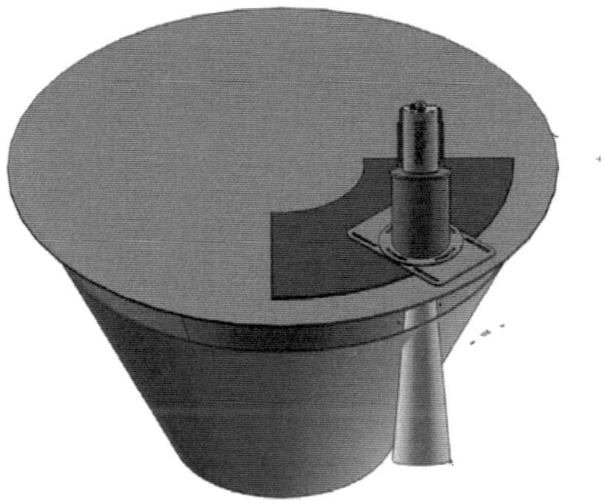

Figure 3 – Schematic of the ColourQ measurement beam on the screen area

As with all instruments used to provide online analytical data, it is essential to maintain a valid calibration setup. With the ColourQ colorimeter, once the calibration is established, drift is managed by regular checking against a calibration tile. This is achieved by swinging the instrument upwards from of its measurement position and inserting a plug into the resultant hole. Attached to the plug is metal rod that the mounting bracket rests on, holding the ColourQ colorimeter in a horizontal position. At this point the instrument lens is cleaned, to remove any residual sugar and syrup that may be present. A frame holding the calibration tile is them clipped to the instrument (see Fig 4), which positions the light beam onto the centre of tile. The calibration process can then be undertaken.

Figure 4 – ColourQ calibration setup position

Performance Measurements

The ColourQ colorimeter is calibrated in ICUMSA solution colour units. Sugar discharging from the centrifuge was collected and analysed for ICUMSA solution colour in the laboratory. The solution colour result was then assigned against the associated colorimeter response for that sugar sample. Sugar samples with a wide range of solutions colours were used to generate a sensible calibration set for the raw sugar installation.

Once the calibration had been generated, it was verified with new samples of discharged sugar with ICUMSA solution colours between 1300 and 3400IU to establish its validity. Figure 5 shows the results of that validity check, showing an R2 correlation of 88% being achieved.

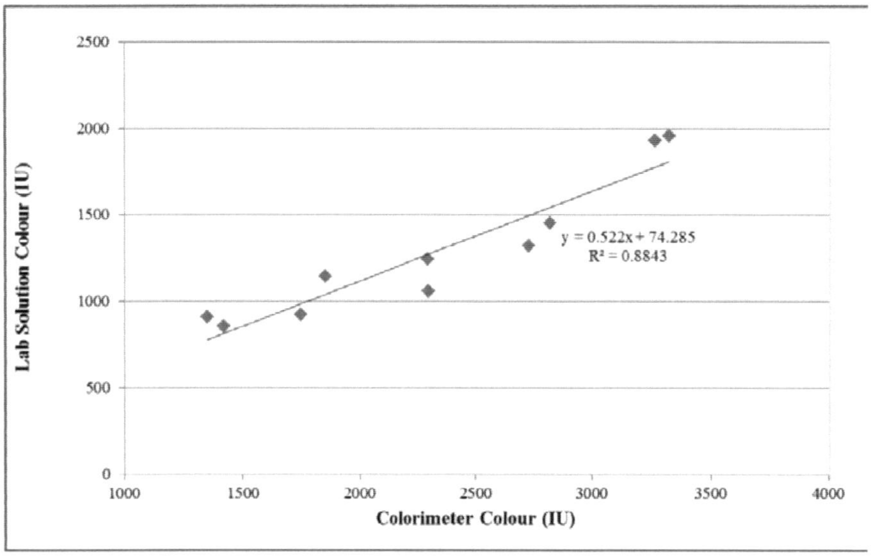

Figure 5 – Colorimeter calibration validation data

Once the team were happy with the performance of the instrument versus laboratory analysis, they started to use the output of the ColourQ colorimeter to evaluate process variance. Variance in discharging sugar solution colour, as measured by the colorimeter, was monitored during normal refinery operations. The operators continued to adjust the wash water ratio on the raw centrifuge as they deemed appropriate for the processing conditions, using both the laboratory analysis and their visual perception of the colour of the sugar discharged from the centrifuge.

Process variance monitoring with the ColourQ colorimeter

Figure 6 below shows changes in centrifuge load (black trace) as operators react to the prevailing conditions in the refinery. The wash water flow (green trace) follows the centrifuge load profile, as the machine was operating in ratio control, managed through the DCS system.

What became apparent is that although the centrifuge was in wash water ratio control, there were times when the solution colour detected by the ColourQ colorimeter (pink trace) changed significantly and this not as would be expected from the relative change in wash water ratio. In

Figure 6 below this is apparent in the first portion of the data set, where the solution colour increases from 700 to 1400 ICUMSA, for a fixed wash water to machine load ratio.

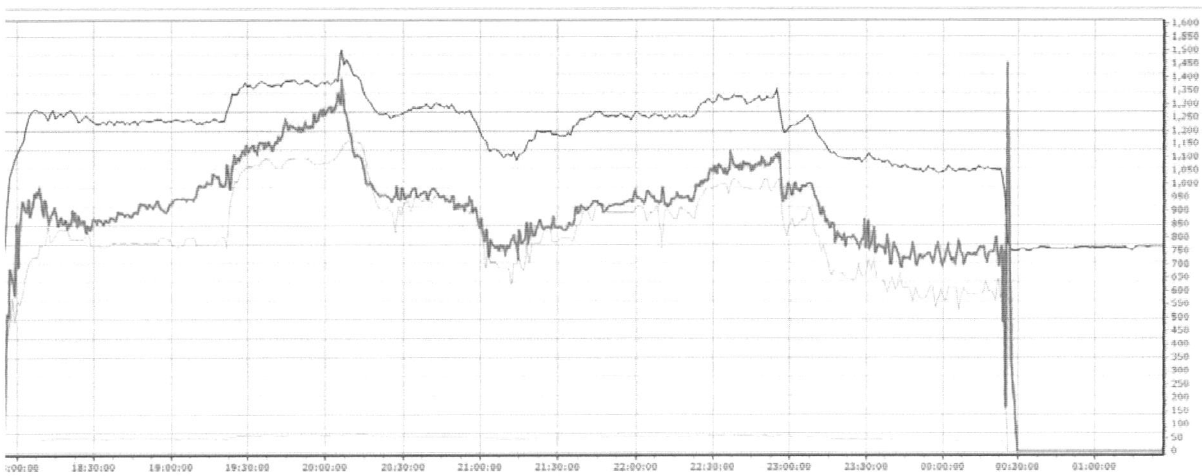

Figure 6 - Typical process variation

This drift in sugar solution colour, for apparently consistent processing condition was a repeating theme and it led the team to investigate the root cause. What they discovered was that as the level of massecuite dropped in the holding receiver before the centrifuge, the solution colour of the discharging sugar from the centrifuge increased, as measured by the ColourQ colorimeter.

Figure 7 illustrates this process variation nicely. The centrifuge load (black trace) and the wash water (green trace) trend up and down together, as defined by their ratio controller, which is altered by the operator. However, the solution colour (pink trace) is varying, even when the load and wash water are not altering and especially when the massecuite receiver (blue trace) is low.

What is also apparent from Figure 7 is the variation within this system. What the team believe they were seeing was variation in massecuite quality from one strike to the next. These changes in massecuite quality were impacting through the continuous centrifuge, and being measured by the ColourQ colorimeter in the solution colour of the discharged sugar. This variation was something noticed throughout the trial, once the solution colour data from the ColourQ colorimeter was available.

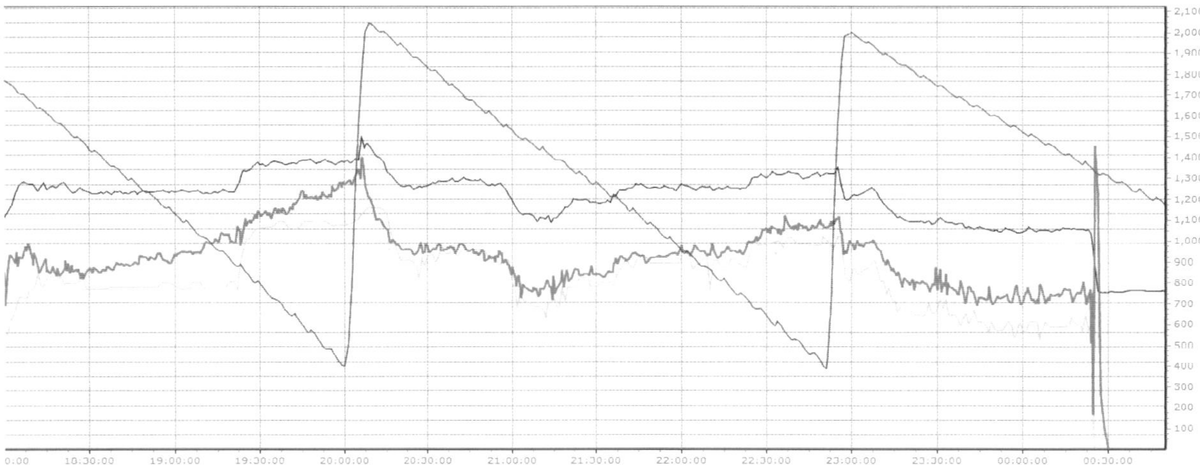

Figure 7 – Process variance between vacuum pans

Development of "Auto Wash" control

Having developed a robust calibration and subsequently used the ColourQ colorimeter to observe the typical process variation in solution colour, the team looked to improve the solution colour control. They wanted to use the ColourQ colorimeter to fully control the addition rate of wash water onto the raw centrifuge. An analogue signal from the ColourQ colorimeter was sent to the factory's DCS system. A piece of DCS logic was built to target the discharging sugar with a solution colour of 1100 IU. As the background process conditions varied the ColourQ colorimeter would detect the changes in solution colour and adjust the wash water addition rate accordingly.

Figure 8 shows how the solution colour (pink trace) is controlling close to the target set point of 1100 IU, once the system was active.

What is also apparent from Figure 8 is how the wash water (green trace) varies with:

- Centrifuge machine load (black trace)
- Massecuite receiver level (blue trace)
- Different strikes

The wash water addition rate (green trace) is much more dynamic and varies more than before. The wash water flow rate is changing between 300 and 700l/hr, where previously this wash water flow would have been constant at a fixed massecuite load.

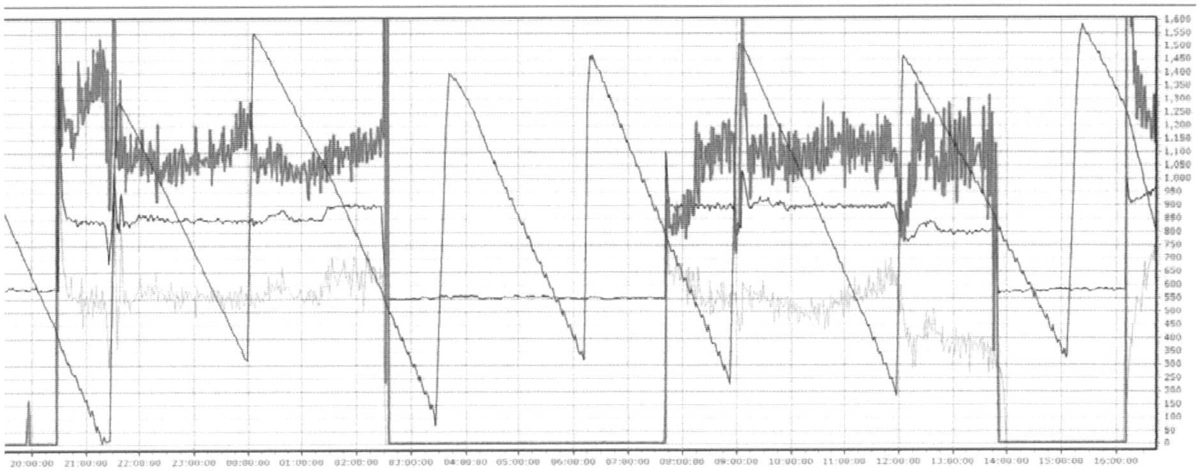

Figure 8 – Process variation with "Auto Wash" enabled

At this point in the project, the reduction in solution colour Standard Deviation was determined. The data analysis showed that the control of solution colour had improved and that it was statistically significant. This gave the team the assurance to raise the solution colour target to 1600 IU (well within normal process variation) from the previous target of 1100IU (see Figure 9).

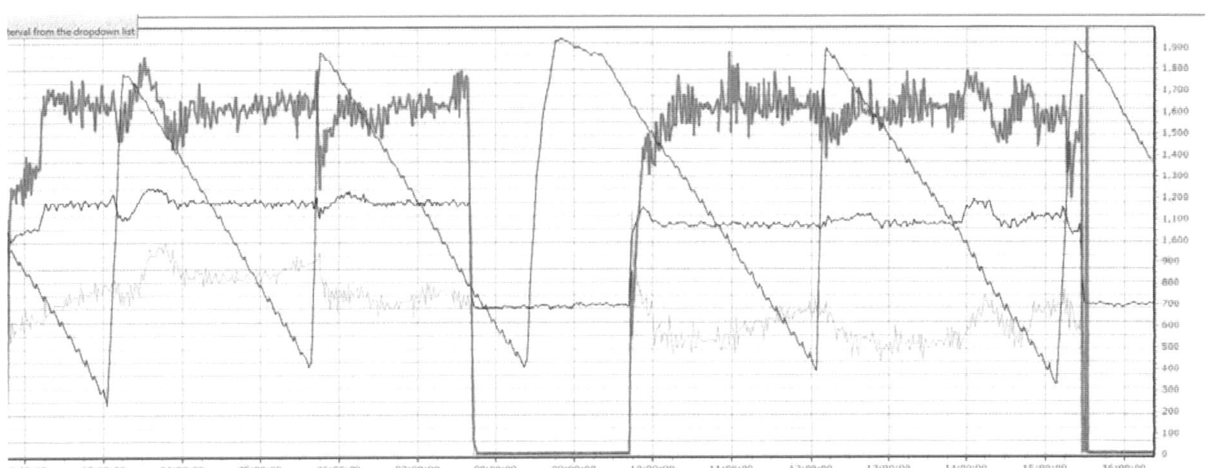

Figure 9 - Process variation with "Auto Wash" enabled and colour target of 1600IU

Results

Reduced colour and impurity recycle

The ColourQ 1700 CC colorimeter has enabled a significant reduction in the variation of solution colour of the sugar being discharged from the raw sugar centrifuge. Being able to have tighter control of the solution colour of the discharging raw sugar will enable an overall reduction in average solution colour of the raw sugar returning to the melter to be realised. This will be due to significantly reducing the number of occasions when the solution colour would rise significantly above the target colour of 1100 IU. This in turn will also reduce the level of impurity recycle with the subsequent positive impacts on overall refinery extraction performance. No extraction benefit (molasses purity reduction) is being reported here due to insufficient analytical data available to statistically demonstrate a benefit above normal process and analytical variance.

However, Figure 10 shows a reduction in solution colour SD (and increase in solution colour) realised with "Auto Wash" enabled and using the solution colour information from the ColourQ colorimeter.

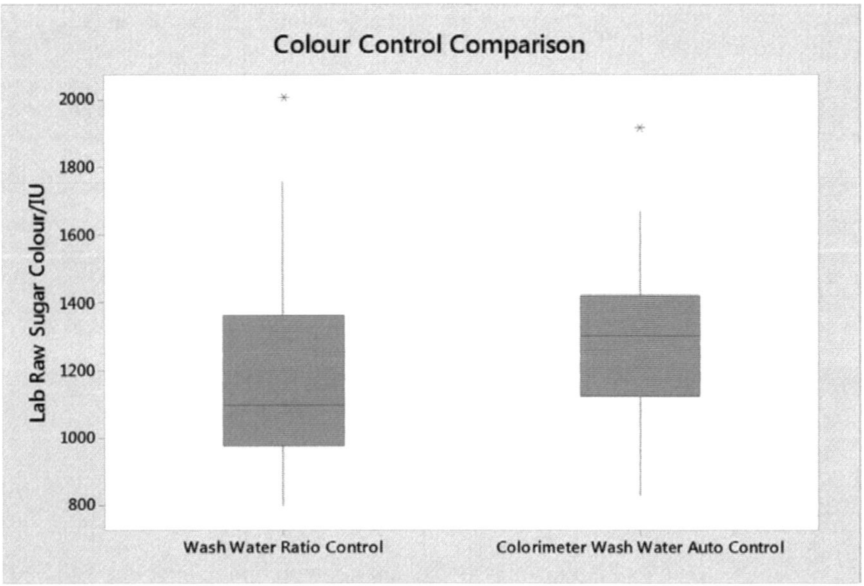

Figure 10 – Increase in Solution colour and reduction in solution colour SD

Improved wash water usage/reduced syrup purity

As described earlier, with confidence in the ColourQ colorimeter's ability to control the solution colour to a much tighter SD, the solution colour target was raised to 1600IU. As a direct result of this change a reduction of 17% in wash water flow was realised. This equates to a dilution (energy) saving of £10,000 a year for one centrifuge. At the same time we also measured a 0.2 reduction in the syrup run off purity from the centrifuge. However, as discussed we have not claimed any benefit for sugar savings.

Figure 11 shows the reduction in wash water usage realised with "Auto Wash" enabled and using the solution colour information from the ColourQ colorimeter.

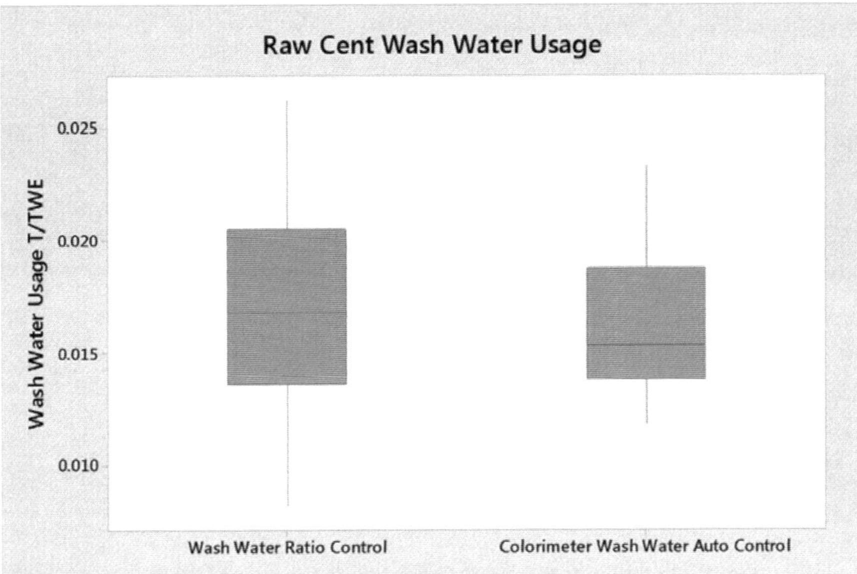

Figure 11 – Reduction in wash water usage

Conclusions

The ColourQ 1700 CC colorimeter has been tested during the Bury St Edmunds' 2017 Juice Refining operation. The instrument has demonstrated that benefits can be realised with either reduced colour/impurity recycle or reduced wash water and syrup run off purity reduction. Additionally, although not part of this trial and as reported by Holcomb, Trent and Nielsen[1], the ColourQ colorimeter can detect "fingers", which are best described as a colour break through the separation line on the centrifuge screen. This leads to additional run off syrup being carried with the discharging crystal back to the melter, leading to a higher level impurity recycle.

Further work is planned with the ColourQ colorimeter, where it will be installed on C-product continuous centrifuge during the 2018 Juice Refining operation. This will allow British Sugar to evaluate direct benefits on molasses purity and any energy benefit around steam and wash water usage.

British Sugar also believes there are a couple of soft benefits that would be realised by installing the ColourQ 1700 CC colorimeter on continuous centrifuges:

A. Reduced laboratory sampling due to the level of accuracy of the ColourQ colorimeter.
B. A significant reduction in the interactions operators would have with continuous centrifuges, freeing up their time.

Acknowledgements

British Sugar wishes to that Bjarne Christian Nielsen (Neltec Denmark A/S) for his support during this trial.

The authors would also like to thank the operational team at British Sugar's Bury St Edmunds factory for providing the resources and support during the trial.

References

1. Optimizing the performance of continuous centrifugals by online monitoring of sugar colour, Trent Holcomb, Tim Diringer, Bjarne Christian Nielsen, Sugar Industry 142, (2017) No.12, p699-703

White Sugar Ash Control via Recovery Sugar Color Control

Connor Bourgeois – Process Engineer, ASR Group

Greg Martin – Process Manager, ASR Group

ASR Group, Chalmette Refinery

7417 N. Peters Street

Chalmette, LA 70043

Abstract

The refinery has evaluated control methods for the ash content of its white sugar. One method that has been examined is controlling the ash concentration of recovery sugars before they are mixed with post-affination liquor in carbonation. Due to the correlation between recovery sugar color and ash concentration, recovery sugar color control can effectively control ash concentration in recovery sugars.

The correlation of recovery sugar color to ash content was measured extensively. All samples were tested for color and ash. After analyzing the data, a correlation was found that closely mimics previously published research. This correlation is used to create a color maximum for recovery sugars to maintain ash specifications in white sugar.

In order to control the color of the recovery sugars, online color measurement is being used for two of the continuous centrifugals in the Chalmette refinery. This instrument sends real time color feedback to the refinery's PLC which then controls the water addition to the continuous centrifugals. A color target is set that allows indirect control of the ash content in the recovery sugars by automatic water adjustments. This influences the concentration of ash that exits carbonation and pressure filtration that ultimately controls the ash concentration of white sugar.

Introduction

About 30%-40% of Chalmette's white sugar business is for customers that require low ash white sugar. In order to satisfy all of the customers' specifications, an ash target of 0.020% has been established for EFG. To achieve this target, the refinery relies on four primary ash-reducing steps: affination separation, carbonation combined with pressure filtration, bone char filtration, and crystallization combined with centrifugation. The optimization of ash removal by recovery syrup crystallization and centrifugation can reduce the amount of ash recycled into the carbonation step of the process.

Recovery-side centrifugation separates crystalline sugar from the refinery's first (High), second (Medium), and third (Final) recovery strikes' massecuites via continuous centrifugals. The recovery massecuite purity targets are 85, 75, and 65 respectively. Screw conveyors transport the sugar from High and Medium strikes to the Remelt Melter while the syrups and the final strikes' sugar are recycled to the recovery vacuum pans to be further exhausted of sucrose. The syrups have a high impurity and ash content relative to the recovery sugar. The impurities and ash are purged from the refinery via the blackstrap molasses stream.

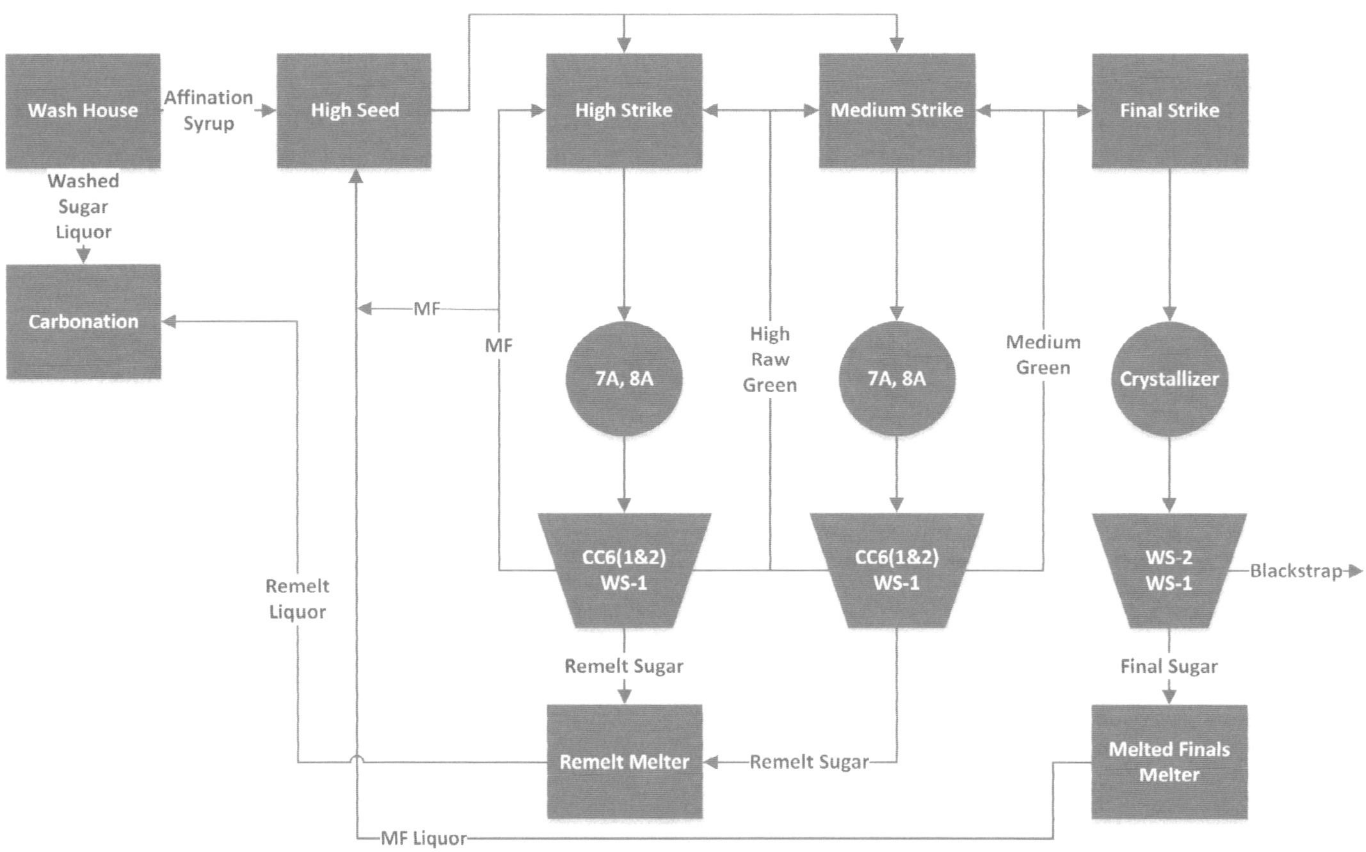

Figure 1: High level Recovery Side process flow diagram for Chalmette

Sugar from the continuous remelt centrifugals is not the only source of ash for the remelt liquor stream, but it is the largest contributor of solids. In addition, most of the other streams that feed this vessel are low in ash content.

Remelt liquor, washed sugar liquor (washed raw sugar dissolved in sweet water), and recycled carbonated liquor are the only liquor streams that feed Chalmette's carbonation vessels. The ash content of the remelt and washed sugar liquor streams can be controlled via centrifugation performance. This control is limited by the amount of ash occluded into the raw sugar and the recovery sugar crystals.

Results

Historical data shows that the best primary ash reduction step is performed by white-side crystallization and centrifugation. In 2017, this unit operation removed 90.6% ± 1.3% (95% confidence intervals) of the ash from its inlet stream. Carbonation and pressure filtration removed 8.62% ± 11.3% (95% confidence intervals) of the ash that it received. Bone char removed 0.15% ± 16.7% (95% confidence intervals) of the ash that it received. Statistically, carbonation and bone char are unreliable at removing ash in Chalmette.

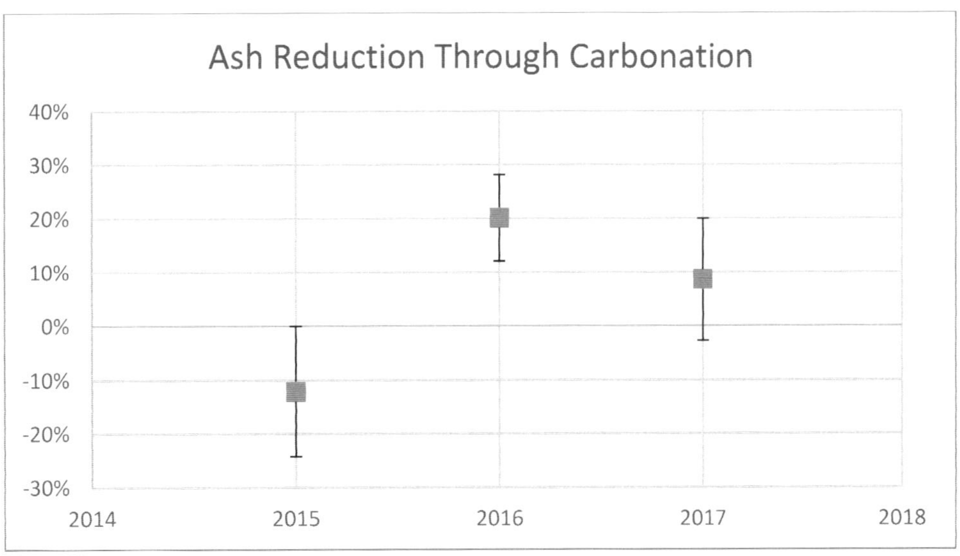

Figure 2: This graph shows the ash reduction through carbonation from 2015 to 2017

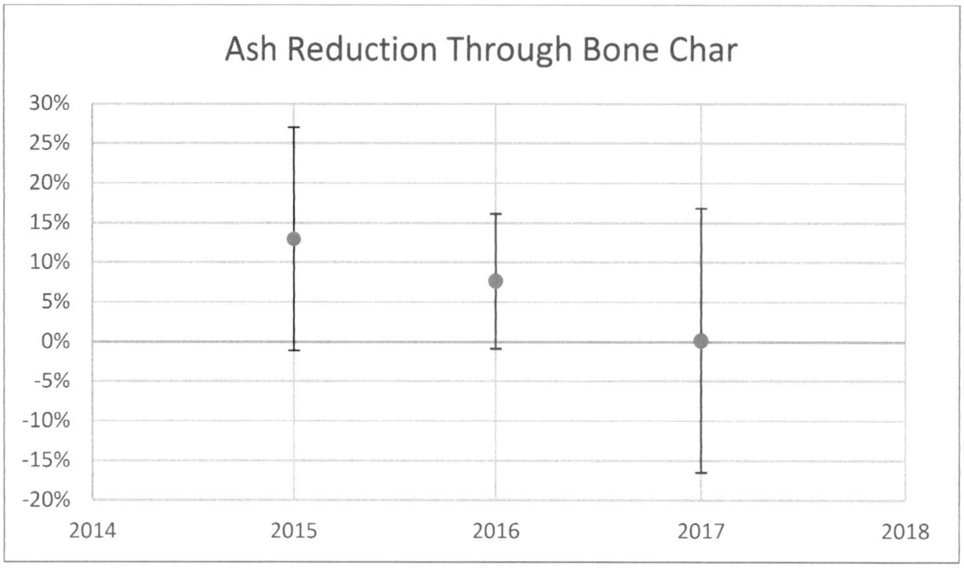

Figure 3: This graph shows the ash reduction through bone char from 2015 to 2017

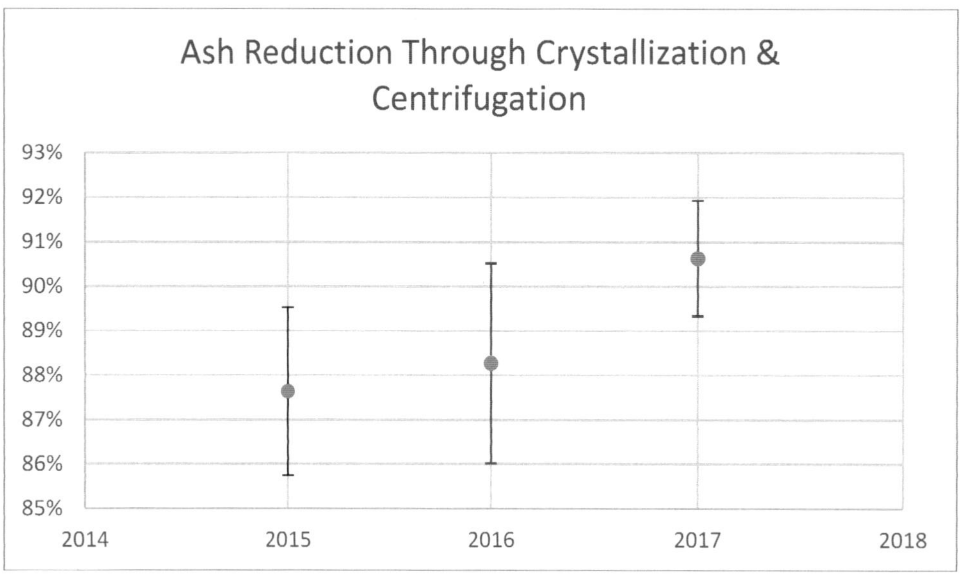

Figure 4: This graph shows the ash reduction through crystallization & centrifugation from 2015 to 2017

The effective ash removal of recovery crystallization and centrifugation matched the performance of white sugar crystallization and centrifugation. The medium or second recovery strikes reduced the ash content of the sugar by 85.5% ± 9.13% (95% confidence intervals). The high or first recovery strikes reduced the ash content of the sugar by 89.2% ± 4.98% (95% confidence intervals).

These average ash removal efficiencies were determined for different unit operations so that an ash target for remelt liquor could be established. An experiment was performed to determine typical ash values in first and second recovery strikes. The average ash content in remelt sugar is

0.44% ± 0.14% (95% confidence intervals). The remelt liquor ash target is defined in equation 1 where WSL is washed sugar liquor.

Equation 1

$$\textbf{Remelt Liquor Ash} = \frac{0.2 - \frac{WSL\ flowrate\ into\ carbonation}{total\ liquor\ flowrate\ into\ carbonation} * (WSL\ ash)}{1 - \frac{WSL\ flowrate\ into\ carbonation}{total\ liquor\ flowrate\ into\ carbonation}}$$

If the actual remelt liquor ash is less than or equal to the requirement found by using Equation 1, the white sugar produced by the refinery will have less than or equal to 0.020% ash content under normal operating conditions. This assumes the washed sugar liquor's ash content is less than 0.2%. If the washed sugar liquor exceeds 0.2% ash, the required color of the remelt sugar becomes too low to attain. At that point, more extraordinary measures (such as increasing wash water use in affination or white sugar batch centrifugals) must be taken to reduce ash content from the primary liquor streams.

To evaluate using recovery sugar ash control to influence white sugar ash control, extensive testing of different grades of recovery sugar and syrups was performed. During normal operation of the recovery pans at the Chalmette refinery, the ash content versus color of the resultant recovery sugar followed a linear trend defined in equation 2.

Equation 2

$$\textbf{Recovery Sugar Ash Content} = \mathbf{0.0001} * \textbf{Recovery Sugar Color}$$

These color values (ICUMSA units) were obtained using an abbreviated refinery color testing procedure where pH is not measured. This relation was established using a substantial dataset with colors ranging from 2,000 to 14,000. The ash analysis was primarily performed via conductivity testing performed by the refinery laboratory. This correlation shows that effective color control is the same as indirect ash control for recovery sugars.

The refinery recently acquired two inline color-measuring instruments for continuous centrifugals. These cameras allow water to be dosed into the machine to control the color of the resultant sugar during normal operation. This also allows the refinery to predict the ash content of the remelt liquor. With this information, the refinery can proactively respond to known ash being recycled into the process.

Based on historical ash values of the washed sugar liquor at Chalmette, default color targets have been established for High, Medium, and Final strikes to control the total ash entering carbonation without negatively affecting the density or purities of the recovery syrups. The current default color targets attempt to maintain the total recovery sugar ash content at 0.40%. If water addition is not enough to control the color of the recovery sugar, throughputs can be modified to further control the colors exiting the continuous centrifugals. Excess recovery centrifugation capacity in a refinery allows for lower color recovery sugars at lower water to massecuite ratios.

Conclusion

Historical ash data allows a refinery to predict ash removal in different unit operations. The primary ash reduction steps in the refining process at Chalmette are crystallization and centrifugation. Carbonation and Bone Char are statistically unreliable at removing ash. To maintain the ash target for white sugar, ash entering carbonation can be controlled to a level that is manageable by white sugar crystallization and centrifugation.

Recovery sugar color control is indirect recovery sugar ash control. A recovery sugar ash target can be set based on expected washed sugar liquor ash content to control the ash of the liquor entering carbonation. The ash at this point in the refining process will determine the ash content of the white sugar due to the consistent performance of white sugar crystallization.

Selection of the Operating Parameters in Sugar Crystallization Control

Lajos Rózsa, Jakab Rózsa
ZUTORA Ltd., Hungary

Seppo Kilpinen, Eero Mielonen
K-PATENTS Oy, Finland

Contact: lajos.rozsa@zutora.com

ABSTRACT

Selection of the operating parameters (setpoints for the variables governing the process of crystallization) is a complex task. In the past it was the pan men's responsibility to determine the value of and control (usually by manually operated valves) vacuum, feed syrup and steam flows based on a very minimum of instruments and experience gained over the years.

One part of the technical literature of the industry has been quite cautious in providing well defined instructions and data regarding the operating conditions to be used during the industrial scale crystallization of sugar; in most cases the views of the "artist" pan-men are reflected in the publications.

The other part of the literature (termed: academic) relies on high level models of the process of crystallization. These are based on material, energy and crystal population balance equations with inevitable simplifications and sometimes questionable results. Their use in actual practice on the pan floor is rather limited.

It is well known that the most important driving force of crystallization is supersaturation. Despite some false claims, there is no single instrument able to measure it directly, because it is a multivariable function of several parameters. With the introduction of the SeedMaster Instruments online calculation of supersaturation plus other important massecuite parameters became reality. Based on the effective use of these instruments a complex equation describing the rate of linear crystal growth was formulated and validated in several sugar mills in pans operating according to normal local practice (L. Rozsa et al.: Crystal growth and crystallization control tactics in industrial sugar crystallizers. Part 1. Crystal growth. ISJ, Oct. 2016). The equation reflects the effects of supersaturation, temperature, non-sugar concentration and crystal content on the rate of crystal growth. The equation provides an excellent tool for the simulation of complete strikes with different operating parameter selections.

In this paper simulated data are used to compare the effect of different operating parameter selections. It is shown that proper setpoint selection (and control) of the important variables, like supersaturation, temperature and level of the massecuite during a strike can result in considerable savings in the use of energy (due to absolutely no use of water, for example), reduction of the time of crystallization and the amount of recycled sugar, mean crystal size (MA) close to target and improved product yield and crystal size distribution (CV).

Proper selection of the setpoints for the important parameters naturally must be followed by a well selected control strategy to be effective. In this paper a control configuration consisting of 5 PID controllers is introduced. They are used to implement supersaturation-based control of steam flow and vacuum combined with massecuite level control based on its solids content. Some results obtained by their use in mills in different parts of the world are also presented.

Keywords: operating parameters, supersaturation, SeedMaster Instruments, crystal growth rate, advanced control of crystallization.

INTRODUCTION

"To make sugar is one of the most difficult things in the world. And to make it right is next to impossible." [1] Mark Twain was certainly not an authority in sugar manufacturing and quite a lot of water has left the Mississippi since his time. However, research on how "to make it right", or with the fashionable word of our times: how to "optimize" sugar manufacturing is still on the list of researchers and plant managers as well.

Sugar crystallization is a very important part of sugar manufacturing. Papers and handbooks on the subject still refer quite often to the expertise of the pan-men to rely on even today, when it is common to find advanced devices of process automation (instruments, PLC-s, DCS-s) in operation in mills and refineries all over the world. It is still not uncommon to find the same old ideas practiced several decades ago now frozen in the software of these devices.

The authors of this presentation have been active since a long time not only in the development of new and unique instruments (process refractometers and SeedMaster Instruments) to be used in sugar crystallization, but in "on the spot" commissioning of these devices in different countries. Introduction of the SeedMaster Instruments [2], [3], [4] made the use of reliable online data (not available before) on supersaturation (SS for short), the most important parameter of crystallization possible. With the availability of online data on SS the need to provide not only an advice, but a complete and tested control solution on the direct use of these data for the advanced control of crystallization in batch pans was a natural requirement. The proposed control solution proved its worth in thousands of strikes in quite a few countries.

The wealth of acquired online data from pans in normal operation and experience gained during the last few years coupled with the (sometimes contradicting and not validated results) of other published research on the subject provided the basis of a thorough study of crystal growth. The aim was to provide a complex equation capable to describe the effect of the parameters governing the rate of crystal growth.

1. PARAMETERS GOVERNING THE PROCESS OF CRYSTAL GROWTH IN INDUSTRIAL SUGAR CRYSTALLIZERS

Crystal growth has been studied in the laboratories and in vacuum pans in operation since a long time. Due to the complexity of the problem and to the different approaches (from the very basic pan-floor practical to the academic) the findings are many and often contradictory.

In [5] a crystal growth rate (linear) model was formulated and validated by using process data from vacuum pans located in two different countries operating according to local practice. This was made possible due to the development and use of a SeedMaster Instrument able to provide reliable online data on supersaturation (SS) and on some other important massecuite parameters as well. The general form of the equation:

$$\frac{da}{dt} = f0(K) * f1(SS) * f2(T) * f3(NS) * f4(y) \quad \text{Eq. 1}$$

or with its validated individual components:

$$\frac{\boldsymbol{da}}{\boldsymbol{dt}} = \boldsymbol{K} * (\boldsymbol{SS} - \mathbf{1.035}) * \mathbf{0.277} * \boldsymbol{e}^{(0.0186*T)} * \boldsymbol{e}^{\left(-1.75*\frac{NS}{100}\right)} * (\mathbf{1} - \boldsymbol{y})^{2} \quad \text{Eq. 2}$$

a:	crystal size (μm)
da/dt:	crystal growth (linear) rate (μm/min)
K:	overall crystallization growth rate
SS:	supersaturation (--)
T:	temperature (°C)
NS:	non-sugar concentration (g/100 g water)
y:	crystal content (by volume, abs.)

The components of Eq. 1 will be briefly discussed one by one as follows.

1.1 K: Overall crystallization growth rate (f0(K))

Crystal growth depends on quite many parameters, like liquid velocity (v) and width (δ) of the stagnant layer on the crystal surface, crystal size (a), liquid density (ρ) and viscosity (η), coefficient of diffusion (D), temperature (T) and purity (P). Non-sugar composition (Nc) has also a strong effect on crystal growth and shape as well. Some of these parameters (v, δ, ρ, η, D) are functions of temperature T, which makes the numerical evaluation of K difficult.

$$f0(K) = K = \mathrm{f}(\mathrm{v}, \delta, \mathrm{a}, \rho, \eta, \mathrm{D}, \mathrm{T}, \mathrm{P}, \mathrm{Nc}) \quad \text{Eq. 3}$$

1.2 Supersaturation (f1(SS))

It is generally accepted since a long time that supersaturation (SS) is the most important parameter of crystallization. Its definition:

$$SS = \frac{sugar\ in\ solution}{sugar\ needed\ for\ saturation}\ (both\ at\ the\ same\ temperature) \quad \text{Eq. 4}$$

SS is a function of several variables [6], namely:

- syrup / mother liquor concentration (C)
- syrup / mother liquor purity (P)
- temperature (T)
- liquid (feed syrup) quality parameters (m, b, c), which are different for feed syrups of beet or cane origin and should be determined locally

SS is a multiparameter function of several variables of the feed syrup or mother liquor (Eq. 5) and there is no single measurement device able to provide reliable data on it; it can only be calculated by taking into account all of its governing parameters [6]:

$$SS = f(C, P, T, m, b, c) \quad \text{Eq. 5}$$

The effect of SS on crystal growth is expressed by:

$$f1(SS) = (SS - 1.035) \quad \text{Eq. 6}$$

1.3 Temperature (f2(T))

The direct effect of temperature on the crystal growth rate is represented by Eq. 7:

$$f2(T) = 0.277 * e^{(0.0186*T)} \qquad \text{Eq. 7}$$

Temperature has considerable indirect effect on quite a few process parameters, like liquid velocity (v) and width (δ) of the stagnant layer on the crystal surface, liquid density (ρ), viscosity (η), coefficient of diffusion (D), non-sugar concentration (at higher temperature higher liquid concentration is needed to have the same SS) etc.

1.4 Non-sugar concentration (f3(NS))

$$f3(NS) = e^{\left(-1.75*\frac{NS}{100}\right)} \qquad \text{Eq. 8}$$

Non-sugar concentration is changing during a strike and its effect is increasing considerably with low purity feed syrups.

1.5 Crystal content by volume (abs.) (f4(y))

$$f4(y) = (1-y)^2 \qquad \text{Eq. 9}$$

Crystal content, though often neglected, has a very considerable effect on crystal growth. The value of f4(y) can change in the f4(y) = 1.00 to 0.20 (y = 0.0 to 0.55) range.

Notes:
Equations 6 to 9 provide information on the direct effect of the operating parameters one by one. Some of these (temperature, crystal content) and additional parameters, like feed syrup / mother liquor concentration have an indirect effect too, on some important massecuite parameters, like viscosity and consistency (see Eq. 3). These have a major effect on massecuite circulation and heat transfer from the calandria. Exact formulation of these indirect effects is very difficult since they depend on vacuum pan construction (presence, or absence of a stirrer, feed input etc.) and product data (like crystal size and size dispersion) as well. Some of the operating parameters have opposing direct and indirect effects on the process of crystallization: increased operating temperature for example can result in better circulation and heat transfer due to reduced viscosity / consistency and in increased growth rate (if SS is kept constant), but can degrade crystal color, if excessive, and results in a sharp drop of SS (if it is not controlled properly) reducing the crystal growth rate.

2. CONSTRAINTS

When selecting the operating parameters, we are faced with some constraints which must be considered. These constraints are closely correlated

- to the operating parameters, and
- to local equipment and practice (vacuum pan, machinery and instruments used, seeding practice etc.).

2.1 Operating parameters

2.1.1 Supersaturation

There are quite different data on the nucleation limit (SS) in the relevant literature ranging from SS = 1.20 to 1.60. The most common and accepted data are trended in Fig. 1. [7]. The nucleation limits depend on syrup / mother liquor purity and temperature as well. These limits should not be exceeded (unless shock seeding is practiced) all over the strike to avoid unwanted nucleation.

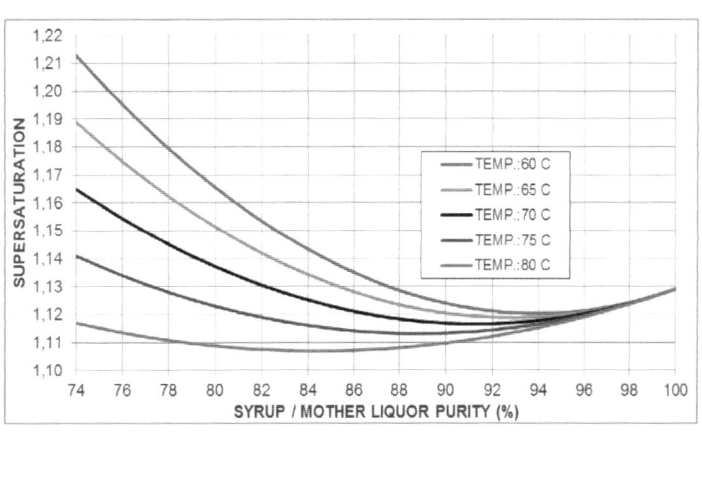

Fig. 1. Critical supersaturation limits to start nucleation [7]

2.1.2 Temperature

Sugar crystallization in vacuum pans is typically carried out in the 62-82 °C temperature range.

2.1.3 Crystal content

Crystal content is strongly correlated to the total solids content of the massecuite. It should be as high as possible when the charge is dropped.

2.2 Constraints posed by local equipment and practice in use

2.2.1 Stirrer motor power rating

There are two parameters which are usually used to signal and start the completion of a strike:

- total solids content or density of the massecuite measured by appropriate instruments (microwave or density probes)
- stirrer motor power or current (power is advised) consumption

Measurement and use of motor consumption data is preferred due to calibration, repeatability and accuracy problems of the microwave and nuclear density probes. Stirrer motors are an important part of the machinery, so they must be protected against overload. However, it happens quite often that the maximum permissible power rating of the motor does not allow continuing the strikes to have higher crystal content at the end of the strikes, though it could be done without any major problem of discharge or centrifuging.
The power rating of the stirrer motor can be a strong limiting constraint on the amount of crystals produced per strike all over the lifetime of the motor (and even longer if the new one used to replace the old will have the same power rating). On the other hand: oversized motors are run well below their specified maximum power rating with higher power consumption (reduced efficiency) all over their lifetime.

2.2.2 Instruments

It follows from Eq. 1 that reliable online data on the parameters having a direct effect on the crystal growth rate (SS, temperature, syrup / mother liquor concentration and non-sugar content, plus crystal content) should be made available for the advanced control of crystallization. The most critical of these is data on SS, which is a must when implementing advanced supersaturation-based control of crystallization. Reliable information on SS is needed not only to implement repeatable seeding, but all over the strike.

Lack of reliable online data on SS is a strong constraint when designing crystallization control.

2.2.3 Control valves

Automatic control of sugar crystallization is based on the use of (among others) control valves to control syrup and steam flows to the vacuum pans, plus whenever possible, to make individual control of vacuum in the pans possible. A few years ago, in a large plant in the Far East, manual control of crystallization was still in use with huge pans of 120 m^3 nominal capacity, partly due to the high cost of the missing large control valves.

2.2.4 Seeding practice

Current seeding practice has two versions:

- full seeding with slurry or footing magma
- shock seeding with slurry

When practicing *full seeding*, supersaturated (SS ~ 1.08) syrup in the pan is seeded by feeding the correct amount of crystals present in the slurry or in the footing magma into the pan. Full seeding is the advised way of seeding.
Shock seeding needs higher supersaturation (above the nucleation limit) in the metastable zone to initiate nucleation by the addition of seed crystals in the slurry (see Fig. 1.).

Lack of reliable online data on SS is a strong constraint in both cases of seeding.

3. INSTRUMENT SELECTION

The set of data needed for the advanced control of crystallization are:

1. Supersaturation
2. Syrup / mother liquor concentration (%)
3. Temperature (°C, °F)
4. Massecuite solids content (%)
5. Syrup / massecuite level (%)
6. Vacuum / absolute pressure (Bar)

3.1 Supersaturation

Feeling the need for data on SS during the process of crystallization, it was the late J.G. Ziegler, pioneer of process automation and its use in sugar crystallization, who had developed a device capable to provide approximate data on SS during a complete strike. Its operation was based on measuring the boiling point elevation (BPE) of the syrup or mother liquor (correlated to its concentration) within the pans under changing temperature conditions [8]. However, BPE of impure sugar solutions is a complex function of liquid concentration, temperature and purity [9]. As it is well known, purity (which is not measured in real time) undertakes a considerable change during crystallization.
With the introduction of the SeedMaster Instruments (developed by Proficon Ltd., ZUTORA Ltd. and K-PATENTS), reliable online data on SS to serve the needs of advanced control of crystallization became reality [2], [3]. In Fig. 2 the third generation of the family (SM-3) is shown.

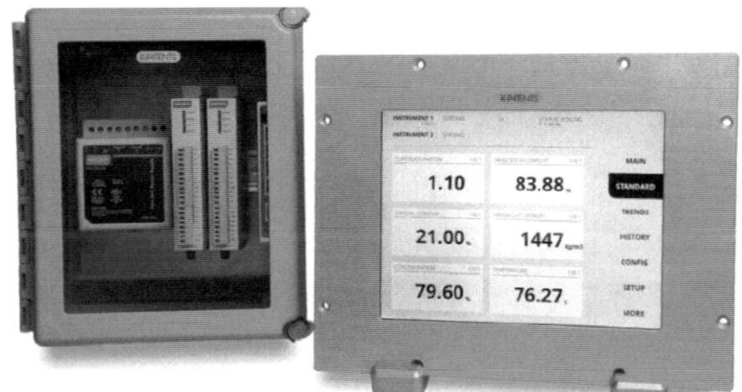

Fig. 2. SM-3 device on the right and the optional I/O unit on the left

Operation of the SM-3 instrument is based on the following information:

- Inputs:
  - Online (real time) data:
    - liquid concentration and temperature measured by a K-PATENTS process refractometer
    - total solids content or density measured by well calibrated third-party microwave (or nuclear) instrument
    - massecuite level measured by a third-party instrument
  - Laboratory data:
    - feed syrup or liquor purity
    - beet / cane syrup quality parameters m, b, c (typical data are available for use in the device)

Based on these data, SS (see. Eq. 5.) and other important parameters are online calculated and trended by the SM-3.

- Calculated data:
  - supersaturation
  - crystal content by volume
  - mother liquor purity
- Other measured, calculated and trended data:
  - massecuite solids content
  - mother liquor concentration
  - temperature
  - level
  - massecuite density
  - mean crystal size, if there was no nucleation during the strike

Important features:

1. A single SM-3 device can serve two vacuum pans simultaneously.
2. Input /output data communication can be implemented via fieldbus communication with a PLC or DCS (use of the hardware I/O unit is optional).
3. Automatic seeding (opening the seed valve for a selected time interval) by the SM-3 is possible (this is the only control action provided by the device).
4. All of the trended data are stored for up to 100 strikes per pan as part of the so-called *Strike History Archive*.

3.2 Syrup / mother liquor concentration and temperature

Reliable and accurate online data on liquid concentration are needed for the calculation of SS (required accuracy is at least +/- 0.1 %). The K-PATENTS process refractometer can meet this requirement and provides temperature data as well (Fig. 3.). No local calibration is required.

Fig. 3. Process refractometer manufactured by K-PATENTS Oy.

3.3 Massecuite solids content (%)

Data on massecuite solids content is provided nowadays by microwave instruments. Their typical accuracy is about +/- 0.2 to 0.5 %. They need local calibration, which is not an easy task. Nuclear density probes, used earlier have lost ground due to several reasons.

3.4 Massecuite level (%) and vacuum / absolute pressure (bar)

Massecuite level measurement by differential pressure transmitters is common practice as well as measurement of vacuum or absolute pressure by pressure transmitters.

4. OPERATING PARAMETERS: A BRIEF OVERVIEW

4.1 Supersaturation

Supersaturation has multiple effects on several important parameters of the massecuite.

4.1.1 Formation of fines and conglomerates due to high SS

Fines and conglomerates provide the basis for recycling already crystallized sugar. Fines can escape through the centrifuge screens, while conglomerates need extensive washing in the centrifuge to improve product color. This operation will dissolve some part of the crystal volume. The recycled dissolved sugar solution will be concentrated and crystallized again, effectively reducing real product yield.

4.1.2 Operational areas of SS during crystallization

The operating range of supersaturation is really small: from SS=1.035 to SS=1.135 (using the nucleation limit SS.lim=1.135 for fairly high purity feed syrups, see Fig. 1.), that is the full range is: ΔSS=0.1 (Fig. 4.). There is absolutely no crystal growth when SS<=1.035.

Supersaturation should be close, but safely below the selected SS.lim all over the time of crystallization. Running a strike way below the safe limit (for example with SS=1.08 instead of SS=1.12) may double the time of crystallization.

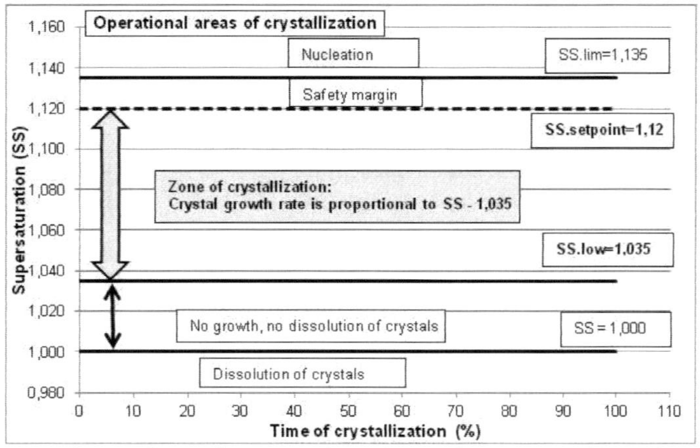

Fig. 4. Operational areas during the course of a strike (seeding practice: full seeding with slurry)

4.1.3 Shock seeding with slurry

Supersaturation must exceed the nucleation limit when shock seeding with slurry is practiced. The number of crystals after completing seeding depends on

- the number of crystals in the slurry,
- the number of new crystals produced and
- the number of crystals dissolved when feeding water to end nucleation.

The number of new crystals produced per minute during nucleation is sharply increasing with increasing SS. The final number of crystals naturally depends on SS and on the length of time spent above the nucleation limit. Trends on supersaturation and crystal content when shock seeding was practiced are shown in Fig. 5. Reproduction of the same SS profile (and so the number of crystals in a unit volume of the massecuite) from strike to strike despite disturbances is not easy. In Fig. 6. trends on SS from 6 consecutive strikes in a refinery in Scandinavia show considerable differences. These certainly had an effect on several important parameters, like the length of time of the strikes, product crystal size (MA) and CV, recycle ratio etc.

Shock seeding practice is in most cases coupled with feeding water to end nucleation. Different publications report on the use of 5 to 10 tons of water per 100 tons of crystal product. Use of water results in increased recycling of sugar and in increased energy consumption.

It is advised to practice full seeding with the correct amount of seed crystals in the slurry, or with well-prepared footing magma.

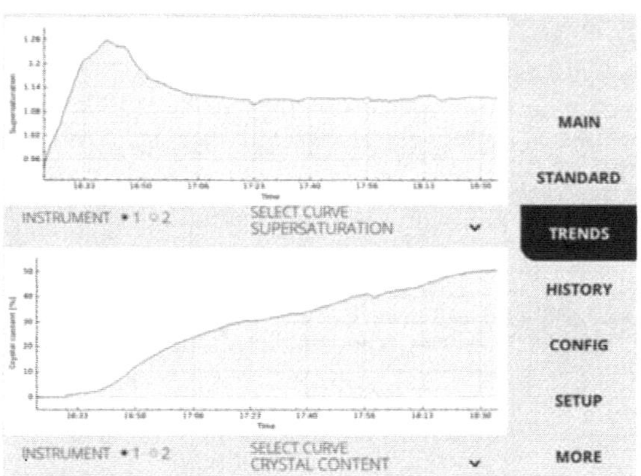

Fig. 5. Trends of SS and crystal content on the SM-3 HMI

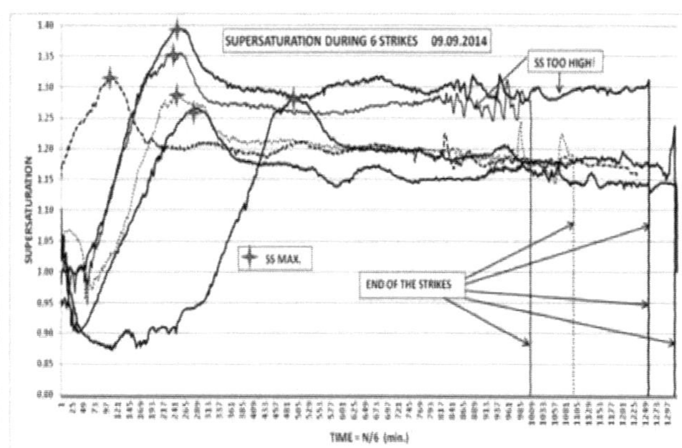

Fig. 6. Trends of SS from 6 consecutive strikes

4.1.4 Inhomogenities in the massecuite due to poor pan circulation

Fig. 7. from [10] shows a typical temperature profile with supersaturation data along the vertical axis of a pan. Temperature is the lowest and supersaturation is the highest on the surface of the massecuite, while the opposite is true at the upper end of the calandria tubes. Too low supersaturation in this latter part of the pan will not only reduce the rate of crystal growth, but can, if it drops occasionally below unity, result in dissolving already crystallised sugar, too.

Uniformity of supersaturation in a vacuum pan depends mostly on the efficiency of circulation and the way feed input is being accomplished and controlled. Central feed (from above or below the calandria) in the downtake results in degraded circulation. Feed should be entered through a ring pipe under the calandria located close to the pan wall.

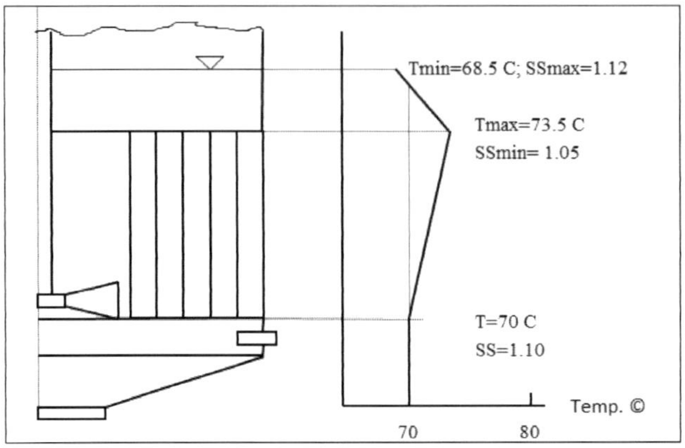

Fig. 7. Typical temperature profile and supersaturation during the first phase of white sugar boiling [10]

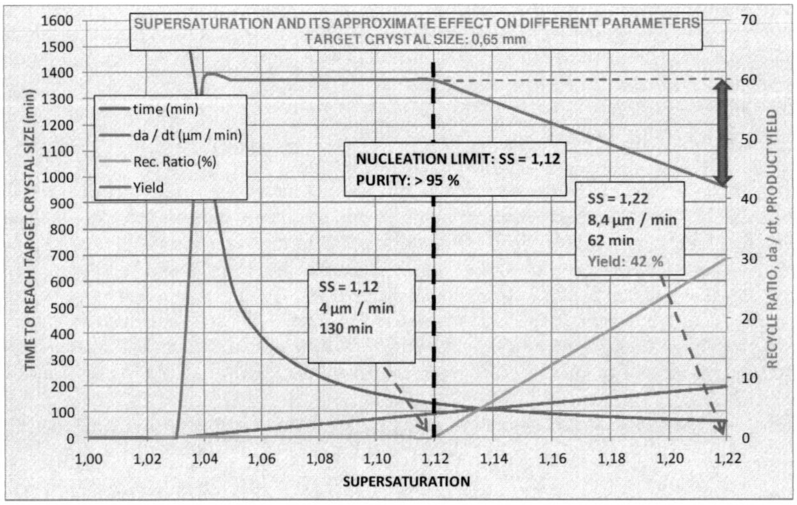

Fig. 8. Supersaturation and its approximate effect on some important parameters

Fig. 8. shows the approximate effect of SS on different strike parameters (time to reach target crystal size, recycle ratio, crystal growth rate and product yield) with feed syrup purities in the P > 95 to 100 % range.

4.2. Operating temperature

Temperature has a profound effect on the process of crystallization.

The operating temperature in vacuum pans during evaporative sugar crystallization typically varies in the 62 to 82 °C range. The direct effect of temperature is expressed by Eq. 7. According to this equation increasing temperature promotes the rate of crystallization.

Some proponents however, argue: it is better to operate crystallization in the low range of temperature, because

- lower pressure steam can be used for evaporation,
- steam consumption is somewhat reduced,
- the danger of color formation is also reduced, and
- mother liquor and non-sugar concentration is smaller.

Use of lower pressure steam: this is not necessarily true since at lower operating temperatures heat transfer from the calandria to the massecuite is decreasing fast.

Color formation: One of the arguments favoring crystallization at low temperatures is lower color formation, which depends on temperature and residence time (time spent at elevated temperature). "White Sugar Boiling: Adequate feed-liquor purity is primary: avoidance of agglomeration, and use of minimum boiling times and temperatures is necessary to avoid (…) invert-color formation" [11]. However, it is clear from Eq. 7. that "minimum temperatures" result in increased boiling times, so these are conflicting requirements (see: Table 1.). The combined effect of temperature and residence time on sugar color was studied [12] and it was found that the increase of color formation begins somewhere above T=73-75 °C.

Smejkal *et al.* [13] discuss the results of experiments conducted in the laboratory with a small crystallizer operated under isothermal conditions. It was found that the "decrease of temperature from 70 °C to 60 °C with SS=1.09 resulted in approximately 30 % decrease of the rate of crystal growth", which agrees with the findings of other researchers.

To help the decision on the selection of the operating temperature during crystallization, calculations were carried out to investigate the role of temperature on some important parameters [14], [15].
The main massecuite data selected for the calculations (same for all operating temperature versions):

Supersaturation:	1.125
Mother liquor purity:	99 %
Syrup m, b, c parameters:	valid for cane syrup in a refinery in Louisiana, USA
Crystal content (y, final):	0.55 vol. abs. (55 % by vol.)
Crystal size (a, final):	0.65 mm

Equations used in the calculations:

Viscosity [16]:

$$VISC = \left(\frac{0.1*(0.99*(14+0.2*C-0.05*T))}{97-6*(0.2*C-0.05*T)}\right)^4 \ (Pa.s) \qquad \text{Eq. 11}$$

VISC:	viscosity of the liquid phase (Pa.s)
C:	liquid concentration (%)
T:	temperature (°C)

Consistency (massecuite) / viscosity (liquid) ratio: R [17]

$$R = \frac{1}{1-y*(3.04*a^{-0.0525}-3.01*y*a^{-0.0965})} \ (-) \qquad \text{Eq. 12}$$

TEMP. °C	CONSIST.rel (%)	f2(T).rel (%)	TIME rel. (%)
72	100	100	100
70	102.0	96.3	103.8
68	104.1	92.8	107.8
66	106.2	89.4	111.9
64	108.4	86.1	116.1
62	110.6	82.9	120.6

Table 1.: Relative change of parameters versus temperature (reference temperature: 72 °C)

The main results are shown in Table 1. Massecuite consistency (relative), f2(T) (Eq. 7) and time of crystallization (from seeding to the end) data belonging to temperatures decreasing from 72 °C (reference temperature) to 62 °C are listed. Appropriate data calculated at 72 °C are used as reference (100 %) data. According to these data, running a strike at 62 °C instead of 72 °C results in 10.6 % larger massecuite consistency (degrading massecuite circulation), and 20.6 % longer time of crystallization. This result is smaller than the one reported by Smejkal (30 %). The difference can be attributed to the decrease of the overall crystallization rate (K), which becomes smaller with decreasing temperature.

4.3 Non-sugar concentration

As already discussed, to have the same value of supersaturation while decreasing operating temperature, mother liquor concentration must be decreased too, and this consequently results in decreasing non-sugar concentration as well. This means that when reducing the operating temperature, the reduced non-sugar concentration has some compensating effect regarding the rate of crystal growth (Eq. 8). This effect becomes stronger with decreasing feed syrup purity.

4.4 Crystal content

Crystal content has a profound effect on the rate of crystallization (Table 2.).

y (by vol., abs)	f4(y)
0.00	1.00
0.25	0.56
0.35	0.42
0.55	0.20

Table 2.: The effect of crystal content on crystal growth rate

Crystal content is determined by the volume (number and size) of the crystals in a unit volume (1 cm^3) of massecuite. Assuming no nucleation all over the strike, the number of crystals in a unit volume will be decreasing due to increasing massecuite volume in the pan. This means that the way massecuite volume (level) is being controlled has a direct effect of on crystal content and on the rate of crystal growth as well.

5. SIMULATION STUDIES

It was proved that crystal content has a strong effect on the crystal growth rate (see Table 2). Based on the validated Eq. 2 a series of strike simulations have been carried out to calculate crystal growth from seeding till the end of the strikes by using two crystallization control tactics (Version No.1, Version No.2) with four different setting parameters. The overall crystal growth rate (K in Eq. 1 and Eq. 2) was selected to equal the one (K=0.0182, sampling rate=10 s) determined during the validation of the growth rate equation Eq. 2. [5].

The main parameters of the simulations:

Feed syrup concentration:	70 %
Feed syrup purity:	98 %
Temperature:	70 ºC (constant)
Syrup m, b, c parameters:	valid for cane syrup in a refinery in Louisiana, USA
Pan seeding volume:	18 m^3
Pan maximum volume:	45 m^3
Supersaturation:	1.125 (constant)

Target data:

Crystal size:	0.65 (mm)
Crystal content by volume:	53 %
Massecuite solids content:	90 %

Besides having the same parameters, the two control versions had something in common too:

- After having seeding (full seeding with slurry) completed, massecuite level (volume) was kept constant for t.wait=10 minutes. During this time crystal size has increased to about 80 to 100 μm, so the danger of dissolving crystals when the level started to increase was negligible.
- Massecuite volume (level) was controlled according to its solids content, but with 4 different setting parameters (b=0.5; b=0.7; b=1.0; b=1.5).

Calculation of the setpoint of level was based on massecuite total solids content according to the equations below:

$$x = \frac{S.ma(t) - S.ma(t.wait)}{S.ma(t.end) - S.ma(t.wait)} \quad \text{Eq. 13}$$

S.ma(t):	massecuite solids content at time t
S.ma(t.wait):	massecuite solids content at the end of time t.wait
S.ma(t.end):	massecuite solids content at the end of crystallization

$$z = x^b \quad \text{Eq. 14}$$

b:	level control parameter

With these data the massecuite level or volume setpoint can be calculated:

$$L(t) = L(t.wait) + z * \left(L(t.end) - L(t.wait)\right) \quad \text{Eq. 15}$$

or

$$V(t) = V(t.wait) + z * \left(V(t.end) - V(t.wait)\right) \quad \text{Eq. 16}$$

5.1 Control Version No.1

Characteristic feature:

Target massecuite volume (45 m^3) and solids content (90 %) were reached at the same time (at the end of crystallization, when dropping of the charge begins). For clarity, the first four figures (Figures 9 through 12) show only trends belonging to the two b parameter extremes: b = 0.50 and b = 1.50.

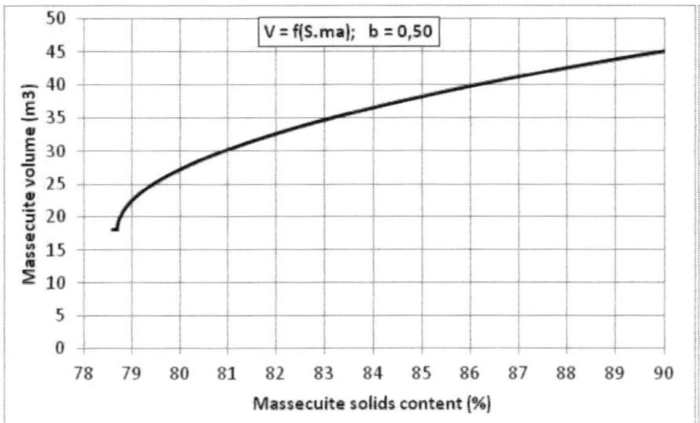

Fig. 9. "Loose" massecuite: volume versus solids content

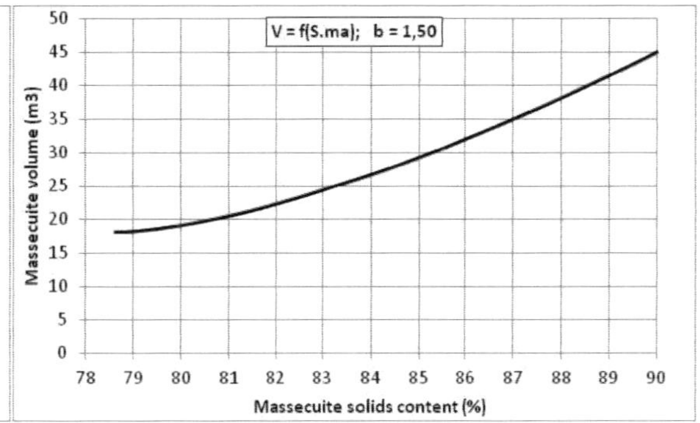

Fig. 10. "Tight" massecuite: volume versus solids content

5.2 Control Version No.2

Characteristic feature:

Target massecuite volume (45 m^3) was reached when solids content was 89 % (1 % less than the target 90 %), then crystallization was continued with constant volume (level) until massecuite solids content reached its target value.

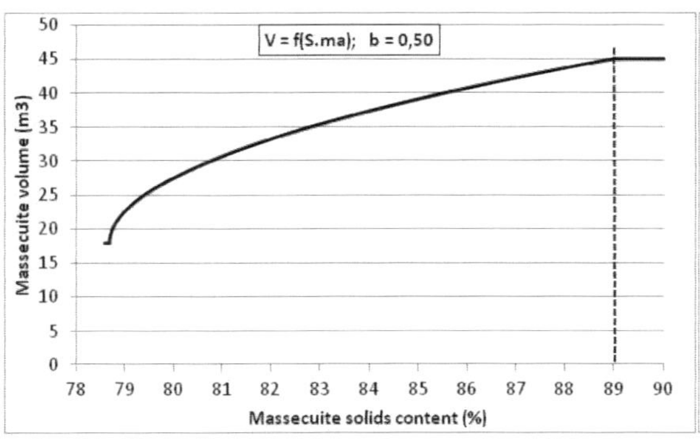

Fig. 11. "Loose" massecuite: volume versus solids content

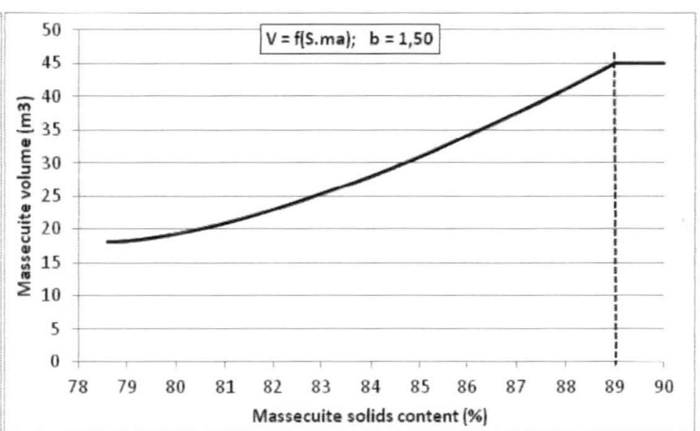

Fig. 12. "Tight" massecuite: volume versus solids content

5.3 Comparison

Comparison of the two control versions is based on some important tabulated data acquired during the simulations.
Symbols used:

b:	parameter used for volume (level) control
a.pr:	product crystal size (mm)
S.ma.end:	massecuite solids content at the end of the strike (%)
V:	massecuite volume (m^3)
V.sug:	total volume of crystals at the end of the strike (m^3)
Time:	time of crystallization from seeding to end of crystallization (min)
Time diff.:	increase of Time (reference: Time with b = 0.50), (min)
Y:	crystal content (by vol., %)
INT. CONSIST.:	time integral of consistency
INT. CONS.rel:	time integral of relative consistency (%) (ref.: b = 0.50)
M. CONS.:	stirrer motor energy consumption (kWh)
M. CONS.rel:	relative motor consumption (%) (ref.: b = 0.50)

b	a.pr (mm)	S.ma.end (%)	V (m^3)	V.sug (m^3)	Time (min)	Time diff. (min)	Y (%)	INT. CONSIST.	INT. CONS.rel	M. CONS. (kWh)	M. CONS.rel
0.50	0.6513	90.01	44.99	23.98	125:30	0:00	53.30	2.14	100.0	61.38	100.00
0.70	0.6515	90.01	45.01	24.00	128:40	+ 3:10	53.34	2.30	107.5	63.71	103.80
1.00	0.6516	90.02	45.01	24.01	132:40	+ 7:10	53.34	2.52	117.5	66.74	108.73
1.5	0.6514	90.01	45.00	23.98	138:00	+ 12:30	53.30	2.84	132.3	71.00	115.67

Table 3.: Version No.1: Summary of the effect of massecuite volume (level) control on some important parameters of crystallization.

As the data in Table 3 show, using "tight" massecuite control instead of "loose" one in the case of *Control Version No. 1* results in

- 12.5 minutes longer strikes,
- 32.3 % larger time integral of consistency meaning considerably poorer massecuite circulation and
- 15.67 % larger stirrer motor energy consumption.

b	a.pr (mm)	S.ma.end (%)	V (m^3)	V.sug (m^3)	Time (min)	Time diff. (min)	Y (%)	INT. CONSIST .	INT. CONS.rel	M. CONS. (kWh)	M. CONS.rel
0.50	0.6512	90.00	45.00	23.97	123:30	0:00	53.26	1.97	100.00	55.89	100.00
0.70	0.6512	90.00	45.00	23.96	125:50	+ 2:20	53.25	2.05	104.29	57.19	102.33
1.00	0.6513	90.00	45.00	23.98	129.00	+ 5:30	53.28	2.18	110.89	59.04	105.64
1.5	0.6513	90.01	45.00	23.98	133:10	+ 9:40	53.30	2.37	120.33	61.56	110.14

Table 4.: Version No.2: Summary of the effect of massecuite volume (level) control on some important parameters of crystallization

In *Control Version No.* 2 using "tight" massecuite control instead of "loose" one (Table 4) results in

- 9.7 minutes longer strikes,
- 20.33 % larger time integral of consistency meaning considerably poorer massecuite circulation and
- 10.14 % larger stirrer motor energy consumption.

As the data show, with Version No.2 and with all variants of parameter b there is an improvement of the important parameters when compared with the similar ones of Version No.1.
Finally: Data in Table 5 and Table 6 help to show even more the differences in two parameters between the two control versions.

b	INT. CONSIST. Ver.2	INT. CONSIST. Ver.1	Ver.1/Ver.2 (%)	Ver.1/Ver2.min (%)
0.50	1.97	2.14	108.87	108.87
0.70	2.05	2.30	112.18	116.99
1.00	2.18	2.52	115.41	127.97
1.50	2.37	2.84	119.72	144.05

Table 5.: Comparing time integrals of massecuite consistencies (Version No.1 and Version No.2)

b	MOT. CONS. Ver.2 (kWh)	MOT. CONS. Ver.1 (kWh)	Ver.1/Ver.2 (%)	Ver.1/Ver2.min (%)
0.50	55.89	61.38	109.82	109.82
0.70	57.19	63.71	111.40	113.99
1.00	59.04	66.74	113.04	119.41
1.50	61.56	71.00	115.33	127.04

Table 6.: Comparing time integrals of motor energy consumptions (Version No.1 and Version No.2)

6. SELECTING THE OPERATING PARAMETERS (SETPOINTS AND SETPOINT PROFILES)

Contribution of the individual components to the rate of crystal growth (see Eq. 1 and Eq. 2) depends on their actual values during the process of crystallization. Some of them are freely selectable for control, like supersaturation (SS) and temperature (T). Others like the overall crystallization growth rate (K), non-sugar concentration (NS) and crystal content (y, abs. by volume) are determined as functions of some other parameters.
Opinions on the optimal values of the operating parameters are many and often contradictory. Some people advise to rely on the expertise and advice of the pan operators. The trajectories of the controlled variables (vacuum, temperature, syrup feed and steam input) are based on experience and on quite many human factors. The way a strike is being controlled might be different even with the same pan operator depending on the time of the day (morning, or night shift).
Handbooks of the industry usually cover this topic briefly and cautiously. Papers on the subject of crystallization are many, and they are mostly based on material, energy and population balance equations used for modeling the process of crystallization. The model data on the different process parameters (SS, crystal content, mother liquor purity, "optimal" steam and feed flow, vacuum and level setpoint profile etc.) are meant to be used for the control of crystallization. While a set of convincing equations are being used to provide the above data, they are all based on quite many simplifications. Calculation of material and energy balances needs much more instruments, including the measurement of feed syrup and steam (plus water, if used) flows and it is general

practice to neglect the problem of instrument accuracy, energy lost to the surroundings, dead-times, time lags and control error. All of this quite often results in surprising (to a trained eye) "optimal" setpoint profiles for example on steam consumption. In one case crystals are growing happily and crystal content is increasing fast with the calculated "optimal" (and assumed as fully accurate) steam flow control even when the calculated SS profile drops well below SS=1.00 for quite a long time.

Based on Eq. 1 and on the discussion presented on the effect of its parameters sugar crystallization in batch vacuum pans should be based on the following operating parameters:

6.1 Operating temperature (T)

Temperature of the massecuite is closely correlated to the absolute pressure (or vacuum) in the pan.
It was proved that operating temperature in the 72 to 75 °C range

- has no detrimental effect on crystal color,
- results in shorter strikes compared with those run at lower temperatures,
- has very positive direct (see: f2(T) in Eq, 7), and indirect effects on quite a few parameters (overall crystallization growth rate K, massecuite circulation, heat transfer, stirrer motor power consumption).

Advised operating temperature: T = 72 to 75 °C

6.2 Supersaturation (SS)

As discussed before, SS is the most important parameter of crystallization. As a general rule, it should be kept safely below, but close to the nucleation limit all over the strike. The actual value of the nucleation limit depends on syrup / mother liquor purity and temperature. The long-time lack of an instrument providing reliable online data on SS in pans in operation and the quite large range of nucleation limits published in the relevant literature make the selection difficult. Based on experience gained using the SeedMaster Instruments in supersaturation-based control of crystallization in quite a few countries:

Advised SS setpoint for seeding: SS=1.08 to 1.10 (full seeding with well-prepared slurry or with footing magma)
Advised SS setpoint after seeding: SS=1.11 to 1.13 all over the strike (feed syrup purity P > 90 %)

These data are in close harmony with the nucleation limit trend data belonging to T = 75 °C in Fig. 1.

6.3 Level (L)

It was proved that fairly "loose" massecuite in contrast to the "tight" one has definite advantages regarding the time of crystallization, massecuite circulation, stirrer motor power consumption, heat transfer from the calandria etc. Having seeded the pan (full seeding with slurry or footing magma):

- Level should be kept constant safely covering the calandria for 8-10 minutes (t.wait).
- After time t.wait level should be controlled based on online data provided by a well calibrated microwave (preferred) or nuclear instrument (see Equations 13, 14, 15).
- The selected level control should provide the same type of benefits as those tabulated in Table 3. (Version 1) or Table 4. (Version 2) with parameter b = 0.5 or b = 0.7. Smaller value of b results in faster syrup feed just after time t.wait, which might result (depending on the type of feed input) in the dissolution of already crystallized sugar.
- "Tightening" (continuing evaporation without feed input) of the massecuite towards the end of the strike should be avoided, because it quite often results in excessive supersaturation provoking nucleation, while losing effective product volume.

Advised value of parameter b = 0.5 or 0.7
Advised control version: Version No. 2.

7. IMPLEMENTING ADVANCED CONTROL BASED ON SUPERSATURATION

One way to implement SS-based control of crystallization is shown in Fig. 13.

Instruments:

1. K-PATENTS process refractometer (provided online data: liquid concentration and temperature)
2. K-PATENTS SM-3 device (provided data: SS, mother liquor purity, crystal content)
3. Third-party microwave or nuclear instrument
4. Third-party level transmitter
5. Third-party vacuum or absolute pressure transmitter

Controllers:

1. PID1: steam input (slave1) controller
2. PID2: SS1 (master1) supersaturation controller
3. PID3: absolute pressure (slave2) controller
4. PID4: SS2 (master2) supersaturation controller
5. PID5: level controller

Operation in brief:
Steam input to the calandria is controlled by the PID1 (slave1) controller operating the steam valve. It has an approximate concentration (liquid) setpoint, which is modified, if needed by the output of its master PID2 (SS1) controller. The two controllers operate in cascade configuration (CASCADE 1).
Absolute pressure (vacuum) in the pan is controlled by the PID3 (slave2) controller operating the vacuum valve. It has its approximate (absolute pressure) setpoint, which is modified, if needed by the output of its master PID4 (SS2) controller. The two controllers operate in cascade configuration (CASCADE 2).
The two master SS controllers have the same SS setpoints.
Level is controlled by the PID5 controller operating the feed valve. Its setpoint is calculated online, as already discussed.

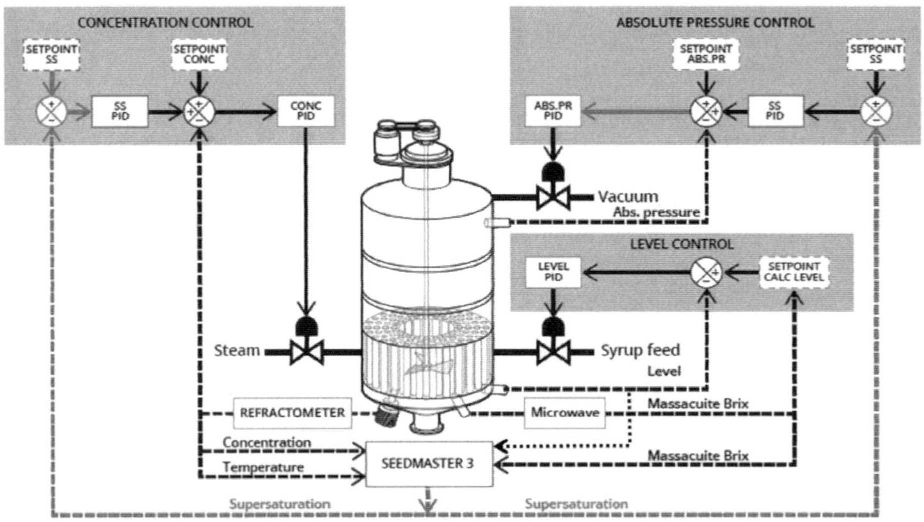

Fig. 13. Configuration of supersaturation-based control of sugar crystallization

The configuration of control shown in Fig. 13. is in operation in several countries of the world.

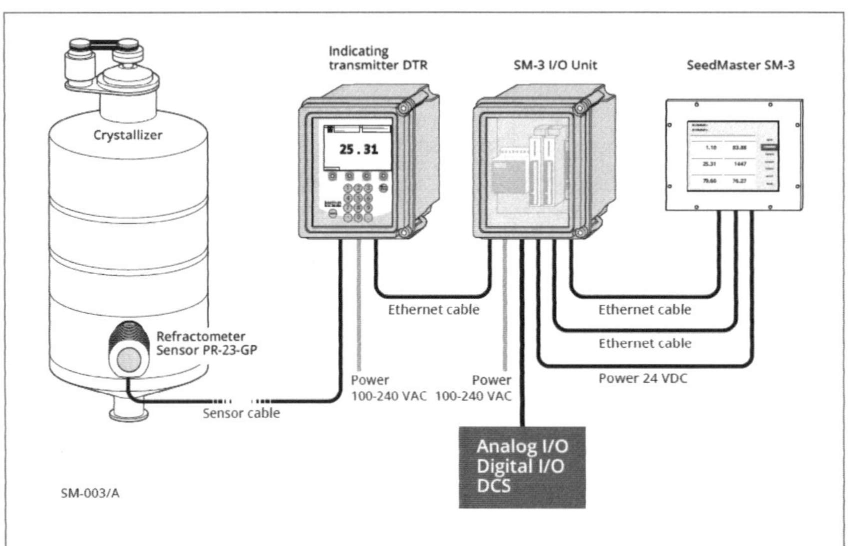

Fig. 14. Process refractometer and SM-3 configuration capable to serve 2 vacuum pans (with 2 sensor heads) simultaneously

The use of the SM-3 I/O Unit shown in Fig. 14. is optional. All input and output data needed for the operation can be transmitted by digital communication between the SM-3 and a PLC or DCS.

8. EXAMPLES AND EXPERIENCE GAINED WITH SUPERSATURATION-BASED CRYSTALLIZATION CONTROL

Data and trends on the important parameters acquired during SS-based control provide a wealth of information which can be used to improve the current practice of crystallization.

8.1 Example No.1

Location: Refinery in West Africa

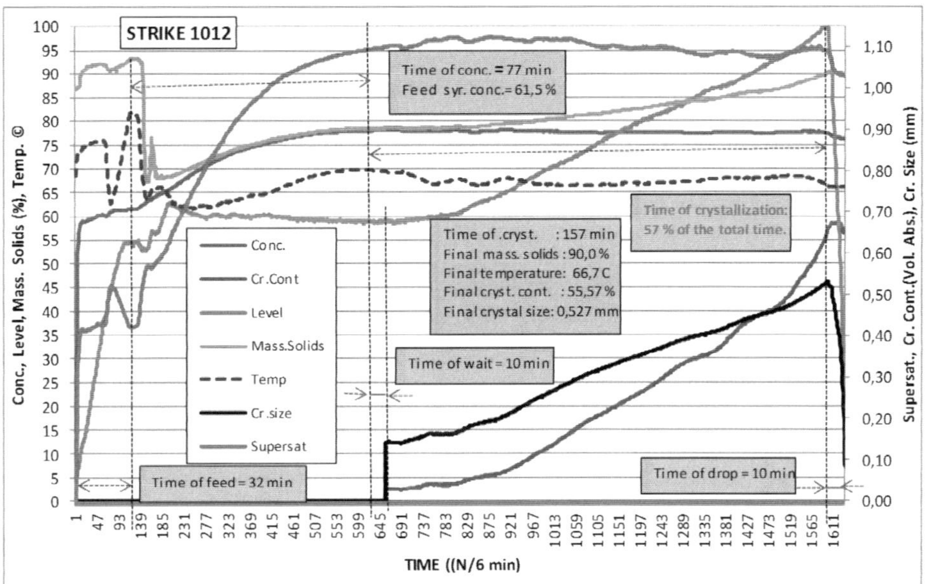

Fig. 15. Trends on the major parameters of a strike

Comments:

1. Supersaturation is fairly well controlled: SS.max=1.125; SS.min=1.075; SS.av=1.102
2. Massecuite level control: Time of wait=10 min; parameter b=1.0
3. Massecuite temperature: Temp.max=69.4; Temp.min=66.4; Temp.av=67.4 (°C)
4. Product crystal content: 57 (% by vol.)
5. Product crystal size: 0.52 (mm)
6. Massecuite solids content: 90 (%)
7. Due to the calibration error of the microwave probe there is a (false) step increase of crystal content and crystal size just after seeding.
8. Running the strike with 5-6 °C higher temperature, a little bit higher supersaturation and smaller b parameter of level control could reduce the time of crystallization effectively.

Notes on local constrains:

1. The actual time of crystallization (from seeding to dropping the charge) is only 57 % of the total time of the strike. This is partly because too much time (32 min) was needed to feed the pan when starting the strike to begin concentration.
2. Syrup concentration to reach seeding takes 77 minutes due to the very low feed syrup concentration (61.5 %). This is the major cause of the small efficiency regarding the use of time defined as time of crystallization over total time of the strike (in this case: 57 %).
3. Due to the very low feed syrup concentration (61.5 %) steam demand during the strike is very high. When two or more strikes are started in similar vacuum pans close in time, serious problems with steam supply may arise and cause difficulties.

8.2 Example No.2

Location: Sugar mill in the Far East

As already mentioned, shock seeding combined with the use of water is practiced in many mills and refineries. A series of tests have been conducted in this plant with the new control method and full seeding with slurry in one of the product pans without using any water during the strikes. The new way of control resulted in considerable improvements regarding the two product crystal parameters, MA and CV (Fig. 16 and Fig. 17). It is well known that use of water after shock seeding dissolves some of the fine crystals and therefore improves the value of CV. Despite this fact, when full seeding with slurry without any use of water all over the strike and the new way of control was introduced, better CV data were achieved.

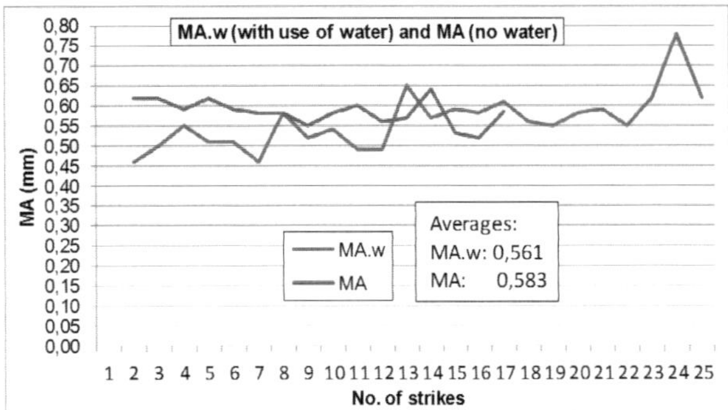

Fig. 16. MA with and without the use of water

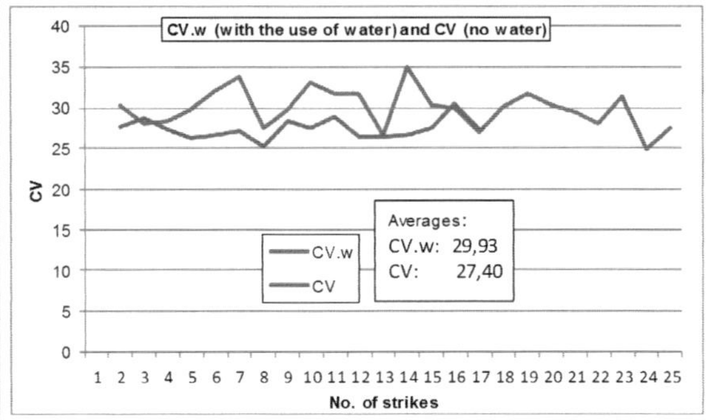

Fig. 17. CV with and without the use of water

If implemented with all of the pans in this plant savings on the use of water and energy could reach about 400.000 USD / year.

8.3 Example No.3

Location: Sugar mill No. 1. in Central America

SUMMARY	SEASON 2015/16	SEASON 2016/17	RESULTS
Crystal Content by Mass (Vacuum Pans)	47,8%	52,9%	11%
Crystal Content by Mass (Centrifugals)	37,9%	44,6%	18%
Dissolution of Sugar by mass (%)	20,8%	15,7%	25%
Sugar Color IU	21,38	24,3	13,7%
Sugar MA	0,46	0,61	33%
Sugar CV	29,43	38,00	29%
Strike Time (min)	168	136	19%
Strike Time best average of 7 days (min)	145	115	21%
No. of strikes per day	16,0	19,2	20%
Max. No. of strikes per day	20	25	25%
Down Time (cleaning included) average per strike (min)	10,6	13,8	30%
Average Ton per Day of Refined Sugar	433	613	42%
Max. Ton per Day of Refined Sugar	536	730	36%

Table 7.: Change of the important production parameters

Table 7. shows comparative data on the most important production parameters collected during two seasons (2015/16 and 2016/17) respectively. The improvements in the second season were partly due to the considerably increased operating temperature and by the introduction of full seeding with slurry. Data with green background reflect improvement, while those with yellow background show degraded data.

Notes:

- Data in the table reflect a massive increase of production in the second season (for example: average ton of refined sugar produced per day was increased by 42 %).
- Sugar color was slightly (+ 13.7 %) increased to 24.3 IU. This was probably due to the reduced dissolution by washing in the centrifugals. It is still much better than the ICUMSA technical specification for white sugar (45 IU max.).
- Sugar MA was increased by 33 % due to changing seeding practice from shock to well-prepared full seeding.
- Sugar CV was increased by 29 %, probably due to the reduced washing in the centrifuges.
- Down time (including discharge and cleaning) was increased probably due to higher crystal content and consistency of the massecuite.

8.4 Example No.4

Location: Sugar mill No. 2. in Central America

In this mill it was common practice to discharge the massecuite when its solids content was 90 % measured by a microwave probe. At that time the stirrer motor load was still below its specified maximum. When the practice was changed to drop the charge a few minutes later when motor load reached its specified maximum, total solids content of the massecuite was 91.7 % and there were no problems encountered due to this change, but the crystal yield per strike was increased accordingly.

Lesson to learn: even small things matter.

9. CONCLUSIONS

Selection of the best values for the major operating parameters has been an important subject of industrial scale sugar crystallization since a long time. Empirical trajectories, products of long-lasting (often never ending) experiments based on the method of trial and error have been commonly used in the industry.

Introduction of the SeedMaster Instruments made the use of reliable online data on supersaturation, never available before, reality. Experience acquired by their use in a large number of strikes in different countries provided the base for the introduction of a crystal growth (linear) rate equation validated with pans in normal operation in two countries [5]. The growth rate equation provided an excellent tool to study the effect of its governing parameters on the process of crystallization.

The importance of supersaturation in crystallization control has long been acknowledged in the industry. Similarly, the quite many effects of temperature and level of the massecuite on the process have been recognized, but due to the complexity of the relationships, opinions on their "optimal" values during a strike are many and sometimes conflicting.

In this paper simulated data are used to compare the effect of different operating parameter selections. It is shown that proper setpoint selection (and control) of the important variables, like supersaturation, temperature and level of the massecuite during a strike can result in considerable savings in the use of energy (due to absolutely no use of water, reduced stirrer motor power consumption and improved circulation), reduction of the time of crystallization and the amount of recycled sugar, mean crystal size (MA) close to target and improved product yield and crystal size distribution (CV).

Proper selection of the setpoints for the important parameters naturally must be followed by a well selected control strategy to be effective. In this paper a control configuration consisting of 5 PID controllers is introduced. They are used to implement supersaturation-based control of steam flow and vacuum combined with massecuite level control based on its solids content. Some results obtained by their use in mills in different parts of the world are also presented.

References:

1. M. Twain: Life on the Mississippi (1883)

2. L. Rozsa: The SeedMaster device for on-line supersaturation measurement and automatic crystallizer seeding
   International Sugar Journal, Vol. C Issue No. 1200, Dec. 1998, pp. 601-607

3. L. Rozsa: SeedMaster 2: A universal crystallization transmitter and automatic seeding device
   International Sugar Journal, Vol. CVIII Issue No.1296, Dec. 2006, pp.683-695

4. L. Rozsa: On-line monitoring and control of supersaturation and other massecuite parameters in vacuum pans: A control engineering approach
   International Sugar Journal, Vol. CXIII Issue No. 1356, Dec. 2011, pp. 853-862

5. L. Rozsa, J. Rozsa, S. Kilpinen: Crystal growth and crystallization control tactics in industrial sugar crystallizers Part 1. Crystal growth
   International Sugar Journal, Oct. 2016

6. L. Rozsa: On-line monitoring of supersaturation in sugar crystallization
   International Sugar Journal, 1996, 98, pp. 660-675

7. R. Broadfoot, P.G. Wright: Nucleation studies
   Proc. 39th Conf. QSSCT, 1972, pp. 353-362

8. J. Ziegler: Sugar Boiling – The Syrups in the Vacuum Pans
   Company Communication

9. M. Saska: Boling point elevation of technical sugar cane solutions and its use in automatic pan boiling
   International Sugar Journal 2002, 104, 1247, pp. 500-507

10. K.E. Austmeyer, T. Frankenfeld: Die radiometrisch erfasste Dichte als Messgröße für den Kochprozess
Zuckerindustrie 110 (1985) Nr. 2.

11. R.A. McGinnis: Beet-sugar technology, Third Edition, 1982

12. F. van Noord, E. Poiesz, A. Wittenberg, M. Ooms: Investigation of cane sugar colour under beet sugar co process conditions
Proc. Sugar Industry Technologists Conference, Auckland, New Zealand, 2012

13. Q. Smejkal, R. Schick, A. Bagherzadeh, L.G. Fleischer: Auswirkungen verringerter Temperaturen bei Eindampfung und Kristallisation auf die Zuckerqualität und den Energiebedarf bei der Zuckergewinnung - Influence of reduced temperatures in evaporation and crystallization on the sugar quality and energy consumption in sugar production
Zuckerindustrie 131 (5/2008)

14. L. Rozsa, J. Rozsa, S. Kilpinen: Crystal growth and crystallization control tactics in industrial sugar crystallizers, Part 2. Control tactics
International Sugar Journal, Sept. 2017

15. L. Rozsa, J. Rozsa, S. Kilpinen: Crystal growth and crystallization control tactics in industrial sugar crystallizers, Part 3. Control tactics (continued)
International Sugar Journal, Oct. 2017

16. S.F. Azevedo, J. Chorao, M.J. Goncalves, L.S. Bento: Monitoring Crystallization Part I, Part II
International Sugar Journal, 1993, 95., (1140): pp. 483-488
International Sugar Journal, 1994, 96., (1141): pp. 18-26

17. A.I. Gromkovskij, V.M. Fursov: Influence of crystal size on massecuiterheology
Izvest. V.U.Z. Pischch. Teknol., 1977, 4, pp. 157-162

Conversion of All VHP Refinery Molasses into Value Added Brown Sugar

Jeffrey Leblanc, PhD, Chemical Engineering,
VP of Technologies, Dr. Chou Technologies, Inc. USA

Jeffrey J Lancon, MBA, P. E., PMP
Principle engineer, Prodigy Engineering Group. LLC, Lafayette. LA. USA

Chung Chi Chou, PhD, Physical Chemistry,
President/principle technologies, Dr. Chou Technologies, Inc. USA

This paper presents processes to increase sugar yield from conventional VHP refineries from 96.5-97.5% to 98.5-99.5% by upgrading molasses to value-added brown sugar. Examples described in this paper have been in plant operation for many years. The example processes include the Areado process, the Muscovado process and the micro-crystallization process. In general, the processes convert massecuite of molasses to molassugar through spontaneous crystallization induced by agitation. Important parameters to control in crystallization processes are the rate of nucleation, the number of nuclei formed, the rate of crystal growth, and the thermal balance between heat of crystallization and latent heat of evaporation. For conversion on final molasses, (1) the massecuite purity should be at least 88 wt. %, (2) the inverted sugars in massecuite should be less than 7%, (3) the massecuite temperature should be above 110 °C, and (4) the massecuite should be at least 94% brix.

In VHP refineries annexed to raw sugar mills or factories, the #2/#3/#4 molasses and syrup streams are recycled back to the mill for further processing, to recover additional sugar during the crop seasons. However, during the "off crop" season, additional boiling cycles to #4/#5/#6 molasses streams are very inefficient, due to extremely high viscosity, resulting in low sugar yield and higher energy costs. A VHP refinery with a fully implemented process for producing value-added brown sugar from refinery molasses would increase sugar yields by up to 2% and reduce energy consumption over traditional refinery schemes. For a VHP refinery with annual production of 200,000 tons sugar the additional revenue could be: 200,000 x 2% x $750/ton = $3 million USD (assuming $750/ton for the value-added products).

Three existing processes produce value-added brown sugar with varying purity and color.

Process		Sugar		
Description	Type	Purity (%)	Color (ICUMSA)	Moisture (%)
Areado	Batch	96.5 - 97.5	3000	2.5
Muscovado	Batch	90.0 - 92.0	8000 - 10000	1.2 - 1.5
Micro-Crystallization	Continuous	94.0 - 96.0	1500 - 2000	0.9

These final products in the form of aggregates range in crystal sizes from 20 to 100 microns and are partially bound to one another by moisture which remains in the final products and/or non-sucrose compounds. Each process will be described in detail in order to explain the technique.

Areado Process

In the Areado process, the first step is to crystallize a sugar solution in a vacuum pan by spontaneous crystallization. The massecuite is heated to 105-110 °C and discharged into a mixer. In the mixer, water is evaporated from massecuite by inducing a vacuum within the mixer, causing evaporation and a subsequent drop in massecuite temperature. The massecuite begins to spontaneously crystallize. The concentrated hot syrup, in the presence of crystals, begins to transform into a wet soft sugar. The soft sugar is then cooled, dried and sieved for packaging.

The RAR Sugar Refinery in Portugal produces about 10,000 tonnes/year of yellow Areado sugar. At RAR, the strike is started by feeding 10-11 m^3 of the mixture of liquor at 75 Brix and 4-5 m^3 of syrup in a 17 m^3 calandria vacuum pan. It is important to remelt all the crystals that may remain in the pan, before starting to boil. This is done by removing the vacuum and allowing the temperature of the feed to increase. After remelting the crystals, the vacuum is increased and boiling starts. After 20 minutes of boiling, the labile zone is reached. Through spontaneous crystallization, the syrup is transformed into a heavy massecuite of small crystals. The massecuite needs to be heated to 105-110 °C and 91-93 brix before being discharged into vacuum crystallizers, referred to as “areadores”.

To avoid sugar degradation, the massecuite must be dropped immediately to the areadores. The areadores are cylindrical closed vessels with an internal paddle type agitator driven at 6 RPM under vacuum. In order to complete massecuite transformation into a soft sugar, it is necessary to achieve a high sucrose crystallization rate and a high water evaporation rate in the areador. The rotating movement created by internal paddle agitator lifts the massecuite, which increases the exposed surface area and improving evaporation rates. Evaporation from massecuite causes a drop in temperature and initiates crystallization. This process will transform the concentrated hot syrup into a mass of wet soft sugar. Final drying the wet soft sugar typically takes 15 minutes.

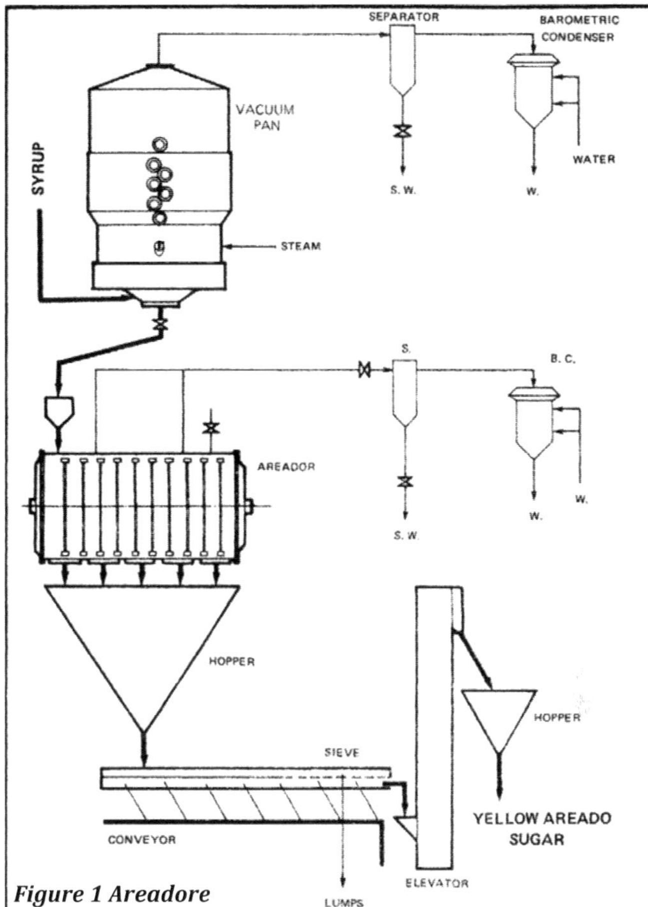

Figure 1 Areadore

The mass and energy balance for the evaporation of could be written as,

$$H_M = H_S + H_V + q$$

$$m_M = m_S + m_V$$

where H_M is the total heat (sensible plus latent heat) of the massecuite, H_S is the total heat of the sugar, H_V is the total heat of vapor, q is the heat loss, m_M is the mass of the massecuite, m_S is the mass of sugar, and m_V is the mass of vapor. Heat released from crystallization will provoke further evaporation and crystallization. At 60 °C the heat of crystallization of sucrose is 57.0 J/g and at 90 °C it is 107.6 J/g. The mass and energy balance are used to determine the working conditions in the areadores. The working conditions determine areado sugar humidity for different purities, massecuite brix and massecuite temperature.

Muscovado Process

Another type of commercial brown sugar is Muscovado sugar. Muscovado is widely produced in many countries but most often in India, Columbia and the Myanmar. Muscovado sugar is also a product of spontaneous crystallization of molasses. Cane juice is boiled at 100°C - 105°C. The resulting massecuite is then vigorously stirred under a flow of air. The lumps are then granulated, sieved and packaged. Muscovado sugar is lower in purity and higher in color than areado sugar.

Micro-crystalline Process

The micro-crystalline process produces brown sugar with slightly lower purity than the Areado process, but having lower color and final moisture content. The process consists of an evaporator, crystallizer, drier, mill and a sieve. The evaporator is operated at a temperature in the range 120 – 130 °C with (or without) vacuum. After evaporation, concentrated syrup of 91 to 97% solids by weight is discharged to the crystallizer. The crystallizer is not under vacuum. Vapors from evaporation are removed from the system through a vapor-liquid separator. Air flows into the crystallizer to remove vapor released during the crystallization. Positive air-flow is produced by a barometric condenser/eductor. The retention time of sugar in the crystallizer ranges from 10-120 seconds depending on the products desired. Specialty sugars with various functionality can be achieved with additives (i.e. maltose, inverts). The functionality can be determined from a nucleation and crystal growth curves. The sugar product leaving the crystallizer has a moisture content between 0.5% and 2.5% by weight. Further drying is applied if needed. The resulting dried aggregates are then cooled. Finally, the sugar aggregates are then milled and sieved.

MolasSugar Process

Process improvements to the Areado process include high torque mixers with increased paddle size for capacity increases and ability to process lower purity process streams. The ability to process lower purity molasses (down to 88 %) would enable the production of three grades of refined sugar. A PID for a modern spontaneous crystallization vessel generated at the request of a refinery with 5,000 ton/day of VPH raw sugar is shown below.

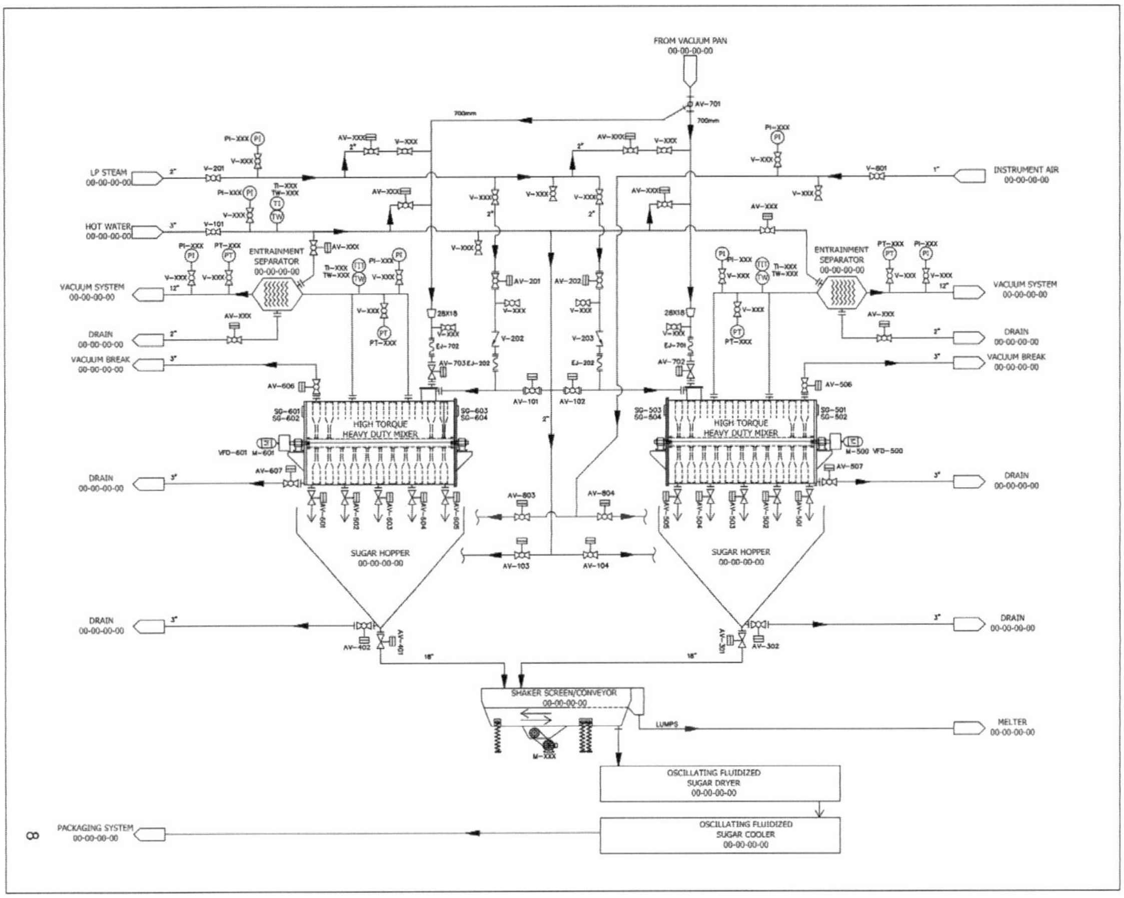

The high torque mixer can be designed for 500 to 1,000 ton/day refinery capacity and would cost between $350,000-$450,000 dollars. The high torque mixer control system is listed in the following table.

Control Loop Description	Method of Control	Equipment Requirement	Control I/O List	Notes
Mixer Speed Control	Monitor Hz from VFD	VFD Drive with Hertz output	Ethernet	
Mixer Drive On/Off	Monitor VFD power	VFD Drive	" "	
Mixer Drive Torque overload	Monitor Amperage of VFD	VFD Drive	" "	
Brix	Monitor Amperage of Mixer Drive Motor	VFD Drive	" "	Indirect method of measuring Brix. Can also use time as a secondary measure.
Mixer's Vacuum Pressure Monitoring	Pressure Transmitter	Pressure Transmitter (PT):	1 AI	Includes separate back-up dial pressure indicator (PI) Pressure control would be part of eductor / condenser control
Entrainment Separator	Pressure Transmitter	Pressure Transmitter (PT)	1 AI	Monitors cleanliness of Entrainment separator. High DP will trigger additional cleaning cycle
Entrainment Separator Cleaning Valves-Open/Close control with Position Indication	Valve Position sensors/relays on valve Pneumatic Actuator	Position sensors on Drain and Water supply valves (Open/Close)	4 DI 2 DO	
Temperature – Monitor	Temperature Transmitter	Temperature Indicator Transmitter (TIT)	1 AI	Not used as a control parameter, monitoring only
Mixer Outlet Valves – Open/Close control with Position Indication (5 loops)	Valve Position sensors/relays on the stem of valve Pneumatic Actuator	Position sensors on Mixer dump valve (Open/Close)	10 DI 5 DO	Each discharge valve will be actuated independently
Vacuum Break control	Valve Position sensors/relays on the stem of valve Pneumatic Actuator	Position sensors on Vacuum Break valve (Open/Close)	2 DI 1 DO	Open close actuation of valve; no flow control.

Cleanout cycle – multiple scenarios ((2) Water Inlet valve, (2) Mixer Drain valves, (1) Steam valve, (1) Air valve)	Valve Position sensors/relays on the stem of valve Pneumatic Actuator	Position sensors on Valves (Open/Close)	12 DI 6 DO	No flow measuring or flow control on water, steam, or air. Control via timers.
Hopper Discharge Valve	Valve Position sensors/relays on the stem of valve Pneumatic Actuator	Position sensors on Valves (Open/Close)	2 DI 1 DO	No flow control. To be operated in conjunction with Cleanout cycle
Condenser / Eductor controls (water flow, Vacuum Pump on/off)	Pump On/Off Controls and status	Relays	2 DI 2 DO	If some control of Vacuum is requested, a VFD can be installed on the cooling water pump.
Condenser / Eductor Water Temperature	Temperature Transmitter	Temperature Indicator Transmitter (TIT):	1 AI	Not used as a control parameter, just monitoring

Poster Title: Radiant heating system for efficient and safe pipe heating

Kapil Kulkarni

Calorique, LLC

2380 Cranberry Highway, West Wareham, MA 02576

Calorique, LLC is a United States based manufacturer located in West Wareham, Massachusetts. Formed in 1980 with roots in traditional residential and commercial radiant heat applications, Calorique, LLC is quickly becoming one of the world's fastest growing companies of advance circuitry with numerous applications and product development. Our unique capabilities and capacity to provide a wide array of heating solutions for many different industries have been the hallmark of the company's success. Calorique serves a wide variety of solutions that cover many industries from home heating and healthcare to commercial aerospace.

As it relates to refinery applications, thin radiant heating films can be used to heat the sugar liquor, dry brown sugar, mesquite and maintain at certain temperatures which help to reduce viscosity and avoid any burnt sugar in the pipelines. By utilizing thin fil technology, it will eliminate the need for any use of open flames to heat up certain sections of the pipes in refineries. The product itself is simple to install and is transferrable between application sites. The product provides a very safe and efficient means of problem solving at the refinery pipeline level.

ROSTER OF DELEGATES - ANNUAL MEETING 2018

Name	Affiliation	Address
Aguirre, Dr. Edgar	Sugar Industry Technologists, Inc.	Myrtle Beach, SC, U.S.A.
Ahlgren, Brad	Calgon Carbon Corporation	Moon Township, PA, U.S.A.
Allaire, Serge	Lantic Inc.	Montreal, CANADA
Allgaier, Greg	Western States Machine Company	Fairfield, OH, U.S.A.
Baker, Richard	ASR Group - Baltimore Refinery	Baltimore, MD, U.S.A.
Banakhar, Aboud S.	United Sugar Company	Jeddah, SAUDI ARABIA
Bar, Daniel	EURODIA INDUSTRIE	Somerset, NJ, U.S.A.
Bavuso, Fernando	Prodek Inc.	Miami, FL, U.S.A.
Bayoumi, Hassan	United Sugar Company of Egypt	Suez, EGYPT
Belcher, Bruce	Graver Technologies	Glasgow, DE, U.S.A.
Bergerhoff, Dr. Peter-Dirk	I.K.B. Industrieplanung GmbH	Pracht, GERMANY
Betts, Herb	K-Patents	Vantas, FINLAND
Bhatt, John	Novasep LLC	Boothwyn, PA, U.S.A.
Bonnet, Susan	Calgon Carbon Corporation	Moon Township, PA, U.S.A.
Bouché, Catherine	ITECA Socadei	Aix in provence, FRANCE
Bourgeois, Connor	ASR Group - Chalmette Refinery	Arabi, LA, U.S.A.
Braun, Niklas - Nordic Sugar AB	Nordic Sugar AB	Arlov, SWEDEN
Brewster, Bill	Buffalo Turbines Inc.	London, CANADA
Brichant, Damien	Novasep LLC	Boothwyn, PA, U.S.A.
Bruhns, Dr. Jürgen	Sugar Industry/Zuckerindustrie	Berlin, GERMANY
Buker, Robert	United States Sugar Corporation	Clewiston, FL, U.S.A.
Burchell, Michael	ASR Group - C & H Sugar Company, Inc.	Crockett, CA, U.S.A.
Burke, Luke	Prosep Ltd.	Roscoe. IL, U.S.A.
Burke, Robert	Prosep Ltd.	Roscoe. IL, U.S.A.
Bryant, Chris	Jord International	Shortland NSW, AUSTRALIA
Canales, Mario M.	Graver Technologies	Glasgow, DE, U.S.A.
Canty Jr., Tod	J.M. Canty	Lockport, NY, U.S.A.
Carbonell, Juan Carlos	ASR Group - C & H Sugar Company, Inc.	Crockett, CA, U.S.A.
Carroll, Doug	Schenck Process LLC	Whitewater, WI, U.S.A.
Carter, George	ASR Group - Redpath Sugar Ltd.	Toronto, CANADA
Castañeda, Jorge	Broadbent Inc.	Ft. Worth, TX, U.S.A.
Charron, Andre	Lantic Inc.	Montreal, CANADA
Chettakul, Sarath	Wellman Co., Ltd.	Bangkok, THAILAND
Chettakul, Sayan	Wellman Co., Ltd.	Bangkok, THAILAND
Chudasama, Arvind	International Sugar Journal	London, ENGLAND
Clarke, Dr. Stephen	Emeritus	Royal Palm Beach, FL, U.S.A.
Cramer, Hans	BMA America, Inc.	Greeley, CO, U.S.A.
Dagallier, Adrien	Gaudfrin S.A.	Saint Germain en Laye, FRANCE
Davidson, Paul	ASR Group - European Division	London, ENGLAND
DeBuvitz, Jon	Putsch & Co., Inc.	Ashville, NC, U.S.A.
Delgado, Idalberto	Graver Technologies	Glasgow, DE, U.S.A.
Demone, David	ASR Group - Yonkers Refinery	Yonkers, NY, U.S.A.
Dibella, Peter	Ramu Agri Industries Ltd.	PAPUA NEW GUINEA
Dionne, Patrick	Lantic Inc.	Montreal, CANADA
Dragobetsky, Lev	Perry Videx LLC	Hainesport, NJ, U.S.A.
Dyess, William	United States Sugar Corporation	Clewiston, FL, U.S.A.
Ehrenhauser, Franz	United States Sugar Corporation	Clewiston, FL, U.S.A.

ROSTER OF DELEGATES - ANNUAL MEETING 2018

Name	Affiliation	Address
El Harti, Amine Abdullah	COSUMAR S.A.	Casablanca, MOROCCO
Elschnig, Lars	BWS-Technologie GmbH	Grevenbroich, GERMANY
Falgout, Dr. Robert	Emeritus	Thibodaux, LA, U.S.A.
Fanning, Mark	Sugar Technology International	Plano, TX, U.S.A.
Freeland, Cindy	Kemira Inc.	Bartow, FL, U.S.A.
Gerards, Mario	RHEWUM GmbH	Remscheid, GERMANY
Gonzalez, Francisco	Graver Technologies	Glasgow, DE, U.S.A.
Goodrow, Fred	American Sugar Refining, Inc.	Arabi, LA, U.S.A.
Gordon, Chris	Imperial Sugar/LDC	Port Wentworth, GA, U.S.A.
Griffin, Donald	Emeritus	Venice, FL, U.S.A.
Gunnerson, Gary H. -	Calgon Carbon Corporation	Moon Township, PA, U.S.A.
Hamilton, Brandon	Calgon Carbon Corporation	Moon Township, PA, U.S.A.
Harriman, Janet	United Sugars Corporation	Edina, MN, U.S.A.
Harrison, Brian	Imperial Sugar/LDC	Port Wentworth, GA, U.S.A.
Hatch, Bob	American Charcoal	Scottsdale, AR, U.S.A.
Hempelmann, Reinhold	BMA AG	Braunschweig, GERMANY
Herrmann, Martin	Putsch GmbH & Co. KG	Hagen, GERMANY
Hill, Christopher	Kemira Inc.	Bartow, FL, U.S.A.
Hirata, Dr. Akio	Mitsui Sugar Co., Ltd.	Tokyo, JAPAN
Horvath, Michael	CSC Sugar LLC	New Canaan, CT, U.S.A.
Ibert, Mathias	ROQUETTE	Lestrem Cedex, FRANCE
Ido, Kyohei	Mitsui Sugar Co., Ltd.	Tokyo, JAPAN
Imamura, Katsuya	Mitsui Sugar Co., Ltd.	Tokyo, JAPAN
Inkson, Dr. Mike	Sugar Knowledge International Limited	Tonbridge, ENGLAND
Ishida, Taiichi	Mitsui Sugar Co., Ltd.	Tokyo, JAPAN
Kaddoury, Abdel Jalil	COSUMAR S.A.	Casablanca, MOROCCO
Kai, Hideaki	Hokkaido Sugar Co., Ltd.	Sapporo, JAPAN
Kilpinen, Seppo	K-Patents OY	Vantas, FINLAND
Kim, In-Deok	Samyang Corporation	Ulsan, KOREA
Kochergin, Vadim	Amalgamated Research LLC	Twin Falls, ID, U.S.A.
Kriedt-Richard, Romney	Sugar Journal	New Orleans, LA, U.S.A.
Kulkarni, Kapil	Calorique, LLC	West Wareham, MA, U.S.A.
Kullman, Kjell	Silver Weibull Sweden AB	Hassleholm, SWEDEN
Lahlou, Mohamed	Putsch GmbH & Co. KG	Hagen, GERMANY
LeBlanc, Dr. Jeff	Dr. Chou Technologies, Inc.	Austin, TX, U.S.A.
LeBlanc, Wilson	Honiron Manufacturing	Jeanerette, LA, U.S.A.
Leedy, Bill	Jacobi Carbons, Inc.	Columbus, OH, U.S.A.
Lofgren, Bob	VEi Global	Ft. Myers, FL, U.S.A.
Martin, Greg	ASR Group - Chalmette Refinery	Arabi, LA, U.S.A.
Martin, Manuel	Putsch GmbH & Co. KG	Hagen, GERMANY
Matsuda, Naoyuki	Mitsui Sugar Co., Ltd.	Tokyo, JAPAN
Mendoza, Juan Jose	Prodek Inc.	Miami, FL, U.S.A.
Mielonen, Eero	K-Patents OY	Vantas, FINLAND
Milne, Timothy	Sugar Technology International	Plano, TX, U.S.A.
Moore, Robert	Purolite	Belvidere, IL, U.S.A.
Moreira, Marcelo	Kemira Inc.	Bartow, FL, U.S.A.
Morales, Alejandro	Florida Crystals Corporation	South Bay, FL, U.S.A.

ROSTER OF DELEGATES - ANNUAL MEETING 2018

Name	Affiliation	Address
Muneef, Abdul Elelah	United Sugar Company of Egypt	Suez, EGYPT
Murray, Bryan	Broadbent Inc.	Ft. Worth, TX, U.S.A.
Mustvedt, Gary	Lantic Inc.	Montreal, CANADA
Myers, Craig	Pixact Ltd.	Tampere, FINLAND
Nadler, Billy	William S. Nadler Jr., Inc.	Belle Rose, LA, U.S.A.
Nagami, Kazuko	Mitsui Sugar Co., Ltd.	Tokyo, JAPAN
Naito, Yoshiyuki	Leybold Co., Ltd.	Tokyo, JAPAN
Nielsen, Bjarne Christian	Neltec Denmark A/S	Vojens, DENMARK
Nolan, Neville	William S. Nadler Jr., Inc.	Belle Rose, LA, U.S.A.
Oh, Kyeong-Seok	Samyang Corporation	Ulsan, KOREA
Parker, Craig	British Sugar PLC	Norfolk, ENGLAND
Parks, Michael	ASR Group - Baltimore Refinery	Baltimore, MD, U.S.A.
Pease, Stephen	Dow Chemical Company	Collegeville, PA, U.S.A.
Pelham, David	United States Sugar Corporation	Clewiston, FL, U.S.A.
Perez, Jose	ASR Group - Chalmette Refinery	Arabi, LA, U.S.A.
Planitario, Sabino	United States Sugar Corporation	Clewiston, FL, U.S.A.
Proano, Angel	Western States Machine Company	Fairfield, OH, U.S.A.
Pron, Frédéric	COMESSA S.A.	Strasbourg, FRANCE
Radinger, Carl Christian	Putsch GmbH & Co. KG	Hagen, GERMANY
Ranade, Bhushan	Jord International	Shortland NSW, AUSTRALIA
Rathke, Thomas	Imperial Sugar/LDC	Port Wentworth, GA, U.S.A.
Rauch, Andreas	SternEnzym GmbH & Co. KG	Ahrensburg, GERMANY
Richard, Dr. Charley	Sugar Processing Research Institute	New Orleans, LA, U.S.A.
Richard, Seth	Louisiana Sugar Refining LLC	Gramercy, LA, U.S.A.
Rozsa, Lajos	K-Patents OY	Vantas, FINLAND
Rua, Claire	ITECA Socadei	Aix-en-Provence, FRANCE
Sabourin, Daryl	ASR Group - Baltimore Refinery	Baltimore, MD, U.S.A.
Sakazaki, Miki	Mitsui Sugar Co., Ltd.	Tokyo, JAPAN
Sarir, Emmanuel	Carbo Solutions International LLC	Los Angeles, CA, U.S.A.
Scherm, Robert	Imperial Sugar/LDC	Port Wentworth, GA, U.S.A.
Schmidt, Hans	BMA- America	Greeley, CO, U.S.A.
Schuetz, Sigurd	RHEWUM GmbH	Remscheid, GERMANY
Schulz, Michael	CSC Sugar LLC	New Canaan, CT, U.S.A.
Schulze, Dr. Bernd-Christoph	ESCON GmbH	Berlin, GERMANY
Seither, Mickey	ASR Group - Chalmette Refinery	Arabi, LA, U.S.A.
Seo, Dae Woo	Haitai Htb	Cheonan-si, KOREA
Sherbondy, John	Jacobi Carbons, Inc.	Columbus, OH, U.S.A.
Sheridan, Michael	Markey R. & Sons Inc.	New York, NY, U.S.A.
Silfversparre, Katarina	Nordic Sugar AB	Arlov, SWEDEN
Simon, Houston	Honiron Manufacturing	Jeanerette, LA, U.S.A.
Singh, G.C.	Emeritus	Chicago, IL, U.S.A.
Sinnard, Robert	Western States Machine Company	Fairfield, OH, U.S.A.
Smith, Neil	United States Sugar Corporation	Clewiston, FL, U.S.A.
Soldan, Jeff	Bauermeister USA	Memphis, TN, U.S.A.
Stark, Thomas	IPRO Industrieprojekt GmbH	Braunschweig, GERMANY
Stohr, Jeff	Western States Machine Company	Fairfield, OH, U.S.A.
Suhr, Mark	Sugar Knowledge International Limited	Tonbridge, ENGLAND

ROSTER OF DELEGATES - ANNUAL MEETING 2018

Name	Affiliation	Address
Takano, Ayumu	Hokkaido Sugar Co., Ltd.	Sapporo, JAPAN
Takase, Kazuyuki	Mitsui Sugar Co., Ltd.	Tokyo, JAPAN
Taylor, Mark	Broadbent Inc.	Ft. Worth, TX, U.S.A.
Theoleyre, Marc-André	EURODIA INDUSTRIE	Pertuis, FRANCE
Thompson, Jack	Louisiana Sugar Refining, LLC	Gramercy, LA, U.S.A.
Topfer, Malcolm	Sugar Knowledge International Limited	Tonbridge, ENGLAND
Tso, Hsi-Chun	Taiwan Sugar Corporation	Tainan, Taiwan, ROC
Uildriks, Roelf	Veco, B.V.	LB Eerbeek, THE NETHERLANDS
van Noord, Frank	Suiker Unie	Dinteloord, THE NETHERLANDS
Vawda, Ahmed	United Sugar Company Saudi Arabia	Jeddah, SAUDI ARABIA
von Horn, Dennis	AMF-Bruns GmbH	Apen, GERMANY
Wassing, Paul	Veco, B.V.	LB Eerbeek, THE NETHERLANDS
Wolf-Wigand, Wolter	Leybold Co., Ltd.	Tokyo, JAPAN
Wywrot, Peter	Buffalo Turbines Inc.	London, CANADA
Yen, Cheng-Wei	Taiwan Sugar Corporation	Tainan, Taiwan, ROC
Yoshikawa, Masakiyo	Hokkaido Sugar Co., Ltd.	Sapporo, JAPAN
Yun, Seok Min	Haitai Htb	Cheonan-si, KOREA

Guests of Delegates - Annual Meeting 2018

Name	Affiliation	Address
Aguirre, Margoth	Sugar Industry Technologists, Inc.	Myrtle Beach, FL, U.S.A.
Braun, Cecilia	Nordic Sugar AB	Arlov, Sweden
Braun, Elin	Nordic Sugar AB	Arlov, Sweden
Braun, Hannes	Nordic Sugar AB	Arlov, Sweden
Braun, Meja	Nordic Sugar AB	Arlov, Sweden
Brühns, Ilsa	Sugar Industry/Zuckerindustrie	Berlin, GERMANY
Clarke, Georgeann	Emeritus	Royal Palm Beach, FL, U.S.A.
Cole, Dr. Marsha	Louisiana Tech University	Ruston, LA, U.S.A.
Demone, Kathy	ASR Group - American Sugar Refining, Inc.	Iselin, NJ, U.S.A.
Dibella, Dianne	Ramu Agri Industries Ltd.	PAPUA NEW GUINEA
Drago, Tatiana	Perry Videx LLC	Hainesport, NJ, U.S.A.
Falgout, Peggy	Emeritus	Thibodaux, LA, U.S.A.
Garcia, Carol	I.K.B. Industrieplanung GmbH	Pracht, GERMANY
Griffin, Maria	Emeritus	Venice, FL, U.S.A.
Harrison, Connie	Imperial Sugar/LDC	Port Wentworth, GA, U.S.A.
Hatch, Karen	American Charcoal	Scottsdale, AR, U.S.A.
Inkson, Win	Sugar Knowledge International Limited	Tonbridge, ENGLAND
Kriedt-Richard, Romney	Sugar Journal	New Orleans, LA, U.S.A.
Larsen, Kirsten K.	Neltec Denmark A/S	Vojens, DENMARK
Milne, Susie	Sugar Technology International	Plano, TX, U.S.A.
Morales, Nivea	Florida Crystals Corporation	South Bay, FL, U.S.A.
Murray, Jackie	Broadbent Inc.	Ft. Worth, TX, U.S.A.
Pelham, Gaye	United States Sugar Corporation	Clewiston, FL, U.S.A.
Richard, Courtney	Louisiana Sugar Refining LLC	Gramercy, LA, U.S.A.
Rodriguez, Tamara	Graver Technologies	Glasgow, DE, U.S.A.
Rodriguez, Tania	Graver Technologies	Glasgow, DE, U.S.A.
Scherm, Andrea	Imperial Sugar/LDC	Port Wentworth, GA, U.S.A.
Seither, Carla	ASR Group Chalmette Refinery	Arabi, LA, U.S.A.
Silfversparre, Gustav	Nordic Sugar AB	Arlov, SWEDEN
Silfversparre, Johanna	Nordic Sugar AB	Arlov, SWEDEN
Silfversparre, Oscar	Nordic Sugar AB	Arlov, SWEDEN
Singh, Anjana	Emeritus	Chicago, IL, U.S.A.
Smith, Laura	United States Sugar Corporation	Clewiston, FL, U.S.A.
Topfer, Dianne	Sugar Knowledge International Ltd.	Tonbridge, ENGLAND
van Noord-Ebshamp, Maartje	Suiker Unie	Dinteloord, THE NETHERLANDS
Vawda, Shereen	United Sugar Company	Jeddah, SAUDI ARABIA
Zhai-Dagallier, Yan	Gaudfrin S.A.	Saint Germain en Laye, FRANCE

Sugar Industry Technologists Inc. (SIT)
4032 Captiva Row
Myrtle Beach, South Carolina 29579
USA

ISBN 978-1-5108-7017-8

9 781510 870178

CURRAN ASSOCIATES INC.
proceedings.com